THE TIME NATURE KEEPS

THE TIME NATURE KEEPS

A VISUAL GUIDE TO
THE CYCLES AND TIME SPANS
OF THE NATURAL WORLD

Helen Pilcher

THE TIME NATURE KEEPS: *A Visual Guide to the Cycles and Time Spans of the Natural World*

Published in North America
by The Experiment, LLC, in 2023.

The Experiment, LLC | 220 East 23rd Street,
Suite 600 | New York, NY 10010-4658
theexperimentpublishing.com

THE EXPERIMENT and its colophon are registered trademarks of The Experiment, LLC. Many of the designations used by manufacturers and sellers to distinguish their products are claimed as trademarks. Where those designations appear in this book and The Experiment was aware of a trademark claim, the designations have been capitalized.

The Experiment's books are available at special discounts when purchased in bulk for premiums and sales promotions as well as for fund-raising or educational use. For details, contact us at info@theexperimentpublishing.com.

Library of Congress Cataloging-in-Publication Data available upon request

ISBN 978-1-61519-952-5
Ebook ISBN 978-1-61519-953-2

Cover design by Jack Dunnington

Conceived, designed, and produced by
UniPress Books Limited
unipressbooks.com

Manufactured in China

First printing April 2023
10 9 8 7 6 5 4 3 2 1

CONTENTS

3 LIFE SPANS

4 GROWTH SPANS

5 BEHAVIORAL SPANS

6 BIOLOGICAL SPANS

INTRODUCTION

IT'S ALL ABOUT TIME

According to advocates of the Big Bang theory, time began with a cataclysmic explosion that occurred around 13.8 billion years ago. It led to the creation of energy and matter and, ultimately, the planet that we live on.

Today, Earth is brimming with life and events that unfold through the lens of time. In this book, we explore the time spans that are occupied by these natural events, in order to find out exactly how nature keeps time. Each chapter represents a different type of time span.

Evolutionary time spans trace planetary-scale narratives, such as the transition of life from the sea onto land, the extinction of species, and the emergence of a primate so powerful it has the capacity to destroy Earth.

Ecological time spans describe the dynamic nature of ecosystems. Beavers, for example, can engineer their classic freshwater ecosystem within weeks, while the decaying remains of a blue whale can create a deep-sea ecosystem that can support life across decades.

Life spans are the space between the start and end of life. An adult mayfly lives for less than a day, yet the oldest living individual tree is more than 4,500 years old.

Growth spans, meanwhile, chart the development that occurs during this time. At one extreme, the Greenland shark takes 150 years to reach sexual maturity; at the other, axolotls retain their juvenile form for life.

Behavioral time spans measure the way that organisms respond to their environment. They can occur across long time periods, such as the migration of the painted lady butterfly, or shorter time frames, like the millisecond traps of the carnivorous bladderwort plant.

Finally, biological time spans are dependent on physiological processes, such as metabolism and hormone production. From the number of breaths taken per minute to the time taken to use the toilet, these intervals chart the innate processes that help organisms to function and survive. Let's explore how nature keeps time.

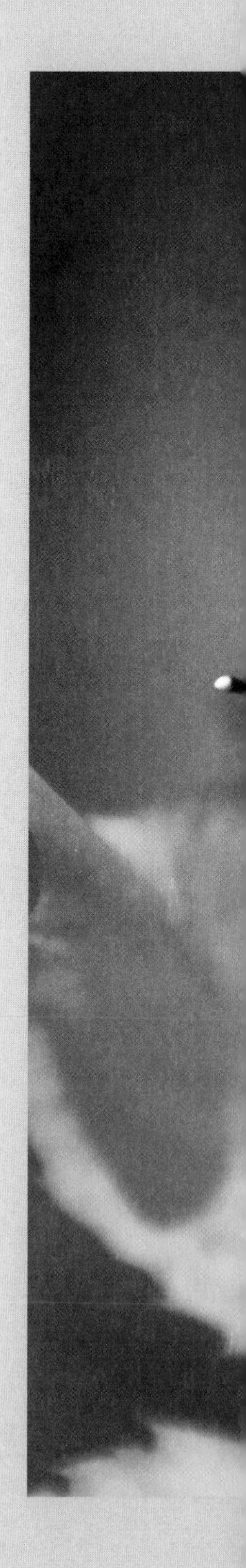

The life of a painted lady butterfly can last just a few weeks, but in that time it can travel thousands of miles.

HAVE A GOOD TIME

From the sweet bouquet of a wildflower meadow to the soulful song of a blackbird at dusk, the natural world has the power to make us feel good. We know this intuitively, but in recent years, scientific studies have lent empirical weight to the idea of a "natural health service" that can benefit us all.

Spending time in green spaces has proven to benefit both physical and mental well-being. People who live in greener neighborhoods have a lower risk of developing cardiovascular disease, obesity, and diabetes. They also report lower levels of stress, better sleep patterns, and higher levels of health and well-being.

The potential benefits of spending time in nature are so great that some doctors now offer "green prescriptions," in which patients are actively encouraged to interact with nature. One program in New Zealand found that after six months, two-thirds of patients issued the prescription felt healthier and

The rise of urban living

Homo sapiens evolve.

First cities are built.

were more active. Eco-therapy, in which people participate in outdoor activities such as gardening or group walks, is also being explored as a possible treatment for some types of depression.

It sounds great, but we all have busy lives. So, what is the optimum "dose" of nature that we should be taking? Once again, science has the answers. A 2019 study of almost 20,000 individuals found the benefits of spending just two hours a week in green spaces is enough to boost feelings of both happiness and health. Whether it is one long walk or several shorter outings, spending time in nature increases physical and mental well-being, and the results apply to everyone, regardless of age, gender, or levels of health and disability.

We are all used to public health messaging telling us to eat five portions of fruit and vegetables per day, or exercise for 150 minutes per week, but perhaps now, it's time to shout about the health benefits of nature, too.

WHY DOES NATURE MAKE US FEEL GOOD?

According to the conservationist Edward O. Wilson, it's because we are part of the natural world. As our species evolved, our brains became primed to respond positively to natural features, such as rivers, forests, and grasslands, that boosted the survival of our ancestors. Proposed in 1984, Wilson called this idea the "biophilia" hypothesis.

10 years ago

More people live in cities than in rural areas.

2050

Seventy percent of people will live in cities.

NATURE'S CLOCKS

From the mechanical clock to the digital wristwatch, humans have been devising increasingly accurate timepieces for centuries, but the natural world has an impressive array of its own timekeepers. Some are tuned to the big picture, recording key events across millions of years, while others are tuned to the small scale, documenting the changes that occur within a single, brief life. They may not have the accuracy of an atomic clock, which has an error of just one second in up to 100 million years, but what they lack in sensitivity, nature's clocks more than make up for in context and color.

The fossil record, for example, tells the dazzling story of life on Earth. Sediments are laid down, layer upon layer, and inside each one are the fossilized remains of some of the organisms that lived during that time period. The study of these layers is called "stratigraphy." It helps paleontologists to determine the relative age of fossils.

HOW TO GAUGE THE AGE OF A FOSSIL

Layers of sedimentary rock are called "strata." The strata at the bottom were deposited first, which means that they are older than the layers above them. Such information can help us to understand

Examples of fossils found in rock

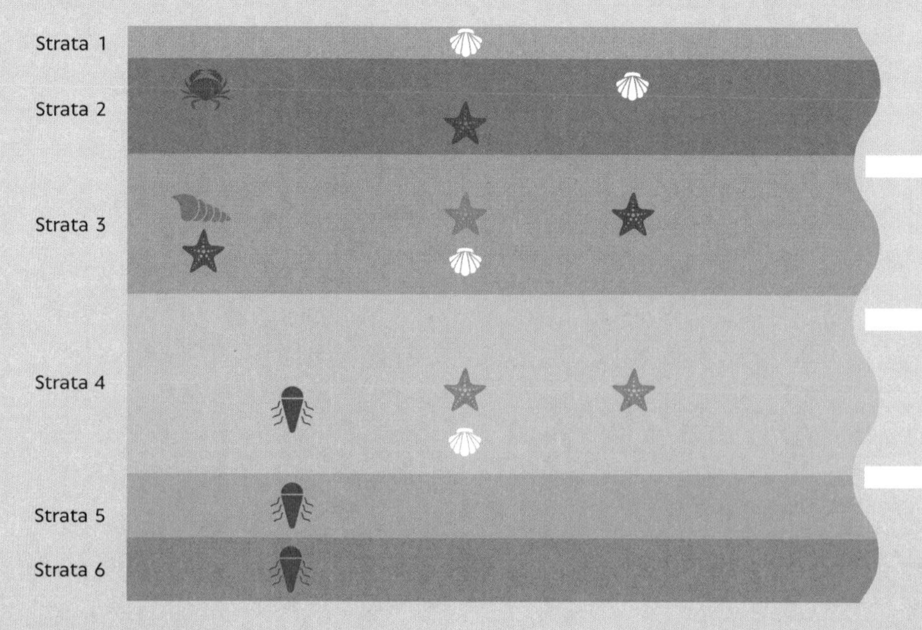

the age of fossils. In the chart below, the gray trilobite found in strata 4, 5, and 6 must be older than the green shell, as the green shell is only located in strata 3. However, the green shell is older than the red crab, which is only found in strata 2.

RADIOMETRIC DATING

Stratigraphy can give us an idea of a fossil's relative age, but to find out its absolute age in "millions of years," other methods are needed. Radiometric dating, which is one of these techniques, relies on the fact that some rocks contain unstable radioactive atoms, which decay at a predictable rate.

Scientists may study radioactive uranium, potassium, or carbon. By measuring the amount of unstable radioactive atoms left in a rock and comparing it to the quantity of stable atoms that have been produced, scientists can estimate the amount of time that has passed since the rock formed.

EXAMPLES OF THE AGE SPAN OF KNOWN FOSSILS

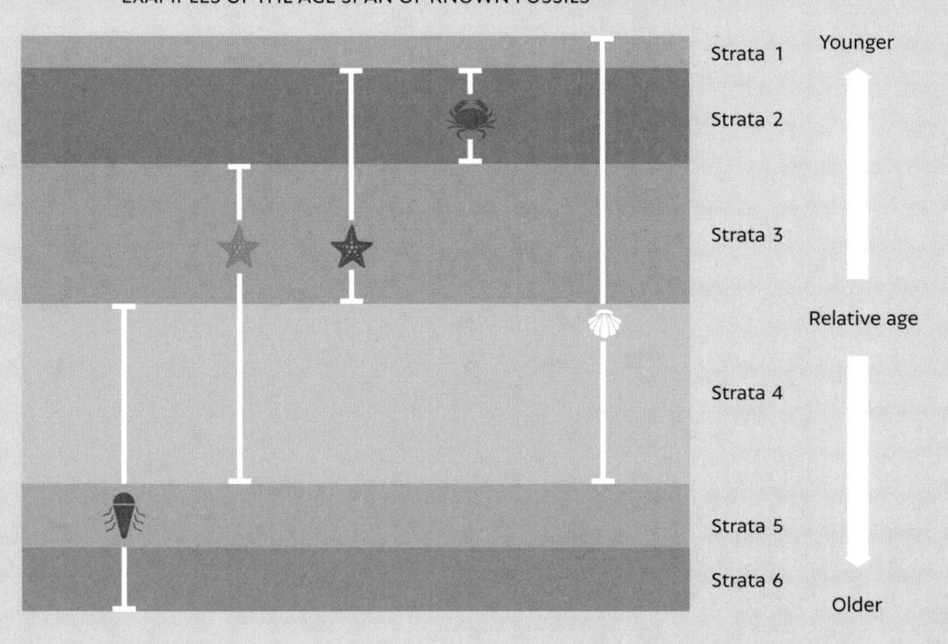

LIVING CLOCKS

Most of us are familiar with the idea that tree rings can be used to measure time. The alternating circles of light and dark give information, not only about the age of the tree but also about the conditions that it experienced through its life.

The same is true for whale earwax. Throughout their life, whales build up layers of lipid-dense wax inside their ear canals. If a whale is found soon after it dies, then the long, thin plugs of earwax can be extracted. When the plug is sawn in half, a series of light and dark bands can be seen, which are counted to give an indication of age; therefore, whale ear plugs can act like a clock.

They can also tell us about the whale's life. Whales often live very long lives, during which they inevitably experience stress. When a whale is stressed, it produces a hormone called "cortisol."

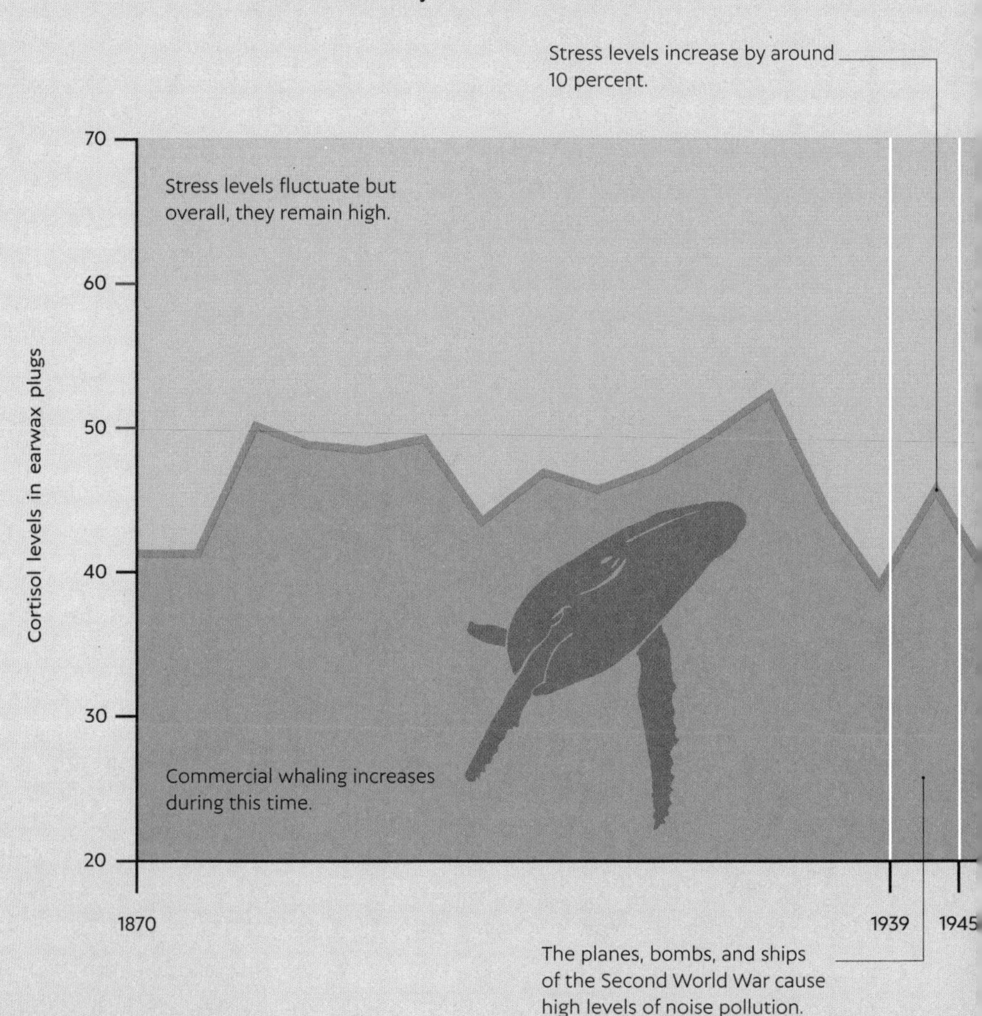

During its lifetime, the levels of cortisol go up and down. These changes are captured in the whale's earwax. Scientists can take samples from different points along the earplugs and then analyze them to identify the cortisol level. Each sample represents a different time in the whale's life. By combining this information, it is possible to tell how the animal's stress levels changed over time.

In a study from 2018, scientists pooled earwax data from 20 whales, spanning over 150 years. Peak cortisol levels, when the animals were really stressed, coincided with periods of negative human activity, such as commercial hunting and war. Now, as the global oceans warm, their stress levels are rising again, so the study shows how detrimental our actions can be to these gentle giants.

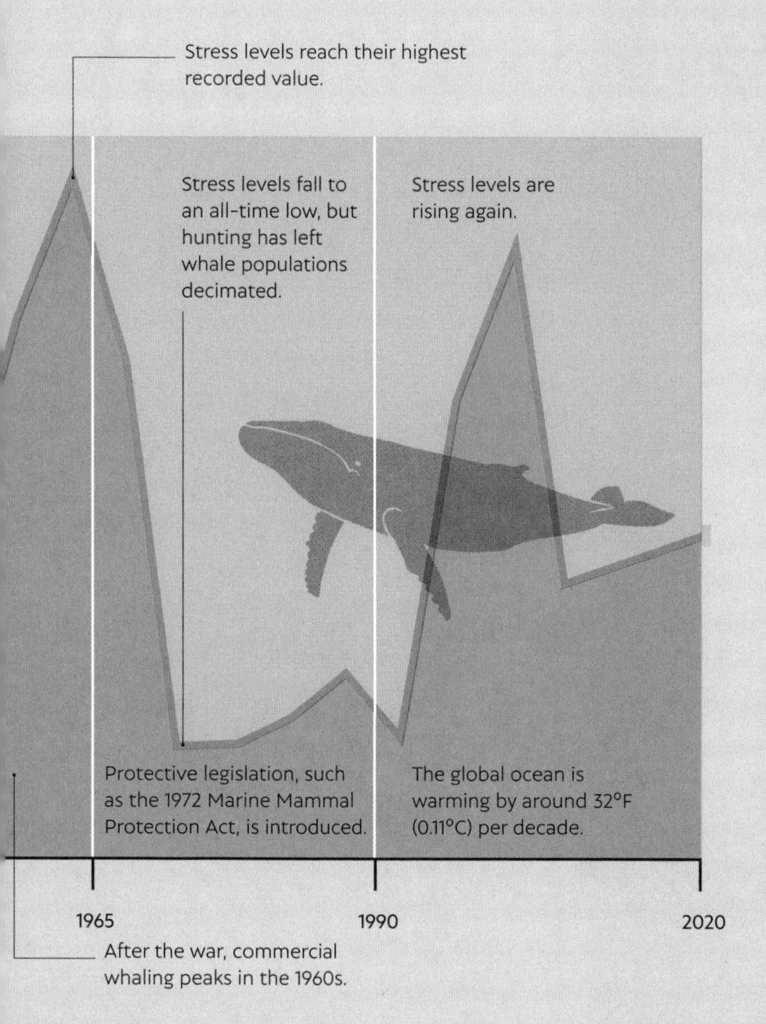

LIFE IN SLOW MOTION

They say that time flies when you're having fun, but we all know how much it can drag when there's housework to be done. The idea that time can seemingly speed up or slow down is familiar to us all, but experiments have revealed that some species can alter their perception of time to suit their needs.

Time perception depends on the speed at which the brain is able to process incoming information. It can be measured by showing animals pulses of light, which start slowly and then speed up. There comes a point where the light is flashing so quickly that it looks like a continuous blur. This is called the "critical flicker fusion frequency." Carefully placed brain electrodes can determine the moment this transition occurs.

Studies show that smaller, speedier animals can detect higher frequencies of flickering lights than larger animals with slower metabolisms. Flies, for example, can perceive light flickering up to four times faster than we can. Just like the fictional character Neo dodging bullets in *The Matrix*, actions and events may seem to unfold more slowly, and while it may explain the infuriating ability of flies to dodge a rolled-up newspaper, it also makes us ask "why?"

From an evolutionary perspective, it makes sense for animals that need to move quickly—to avoid being eaten or to catch fast-moving prey—to perceive time more slowly and react swiftly. It also makes sense for these same animals to be able to change their perception of time when speedy reactions are no longer needed—for example, when resting.

In line with this, scientists have discovered that some swordfish boost blood flow to the brain before they set off hunting. This slows their perception of time and boosts the number of frames they can process per second. When not hunting, time goes by more quickly, as the amount of information they can process per second decreases.

How times moves relative to human time perception

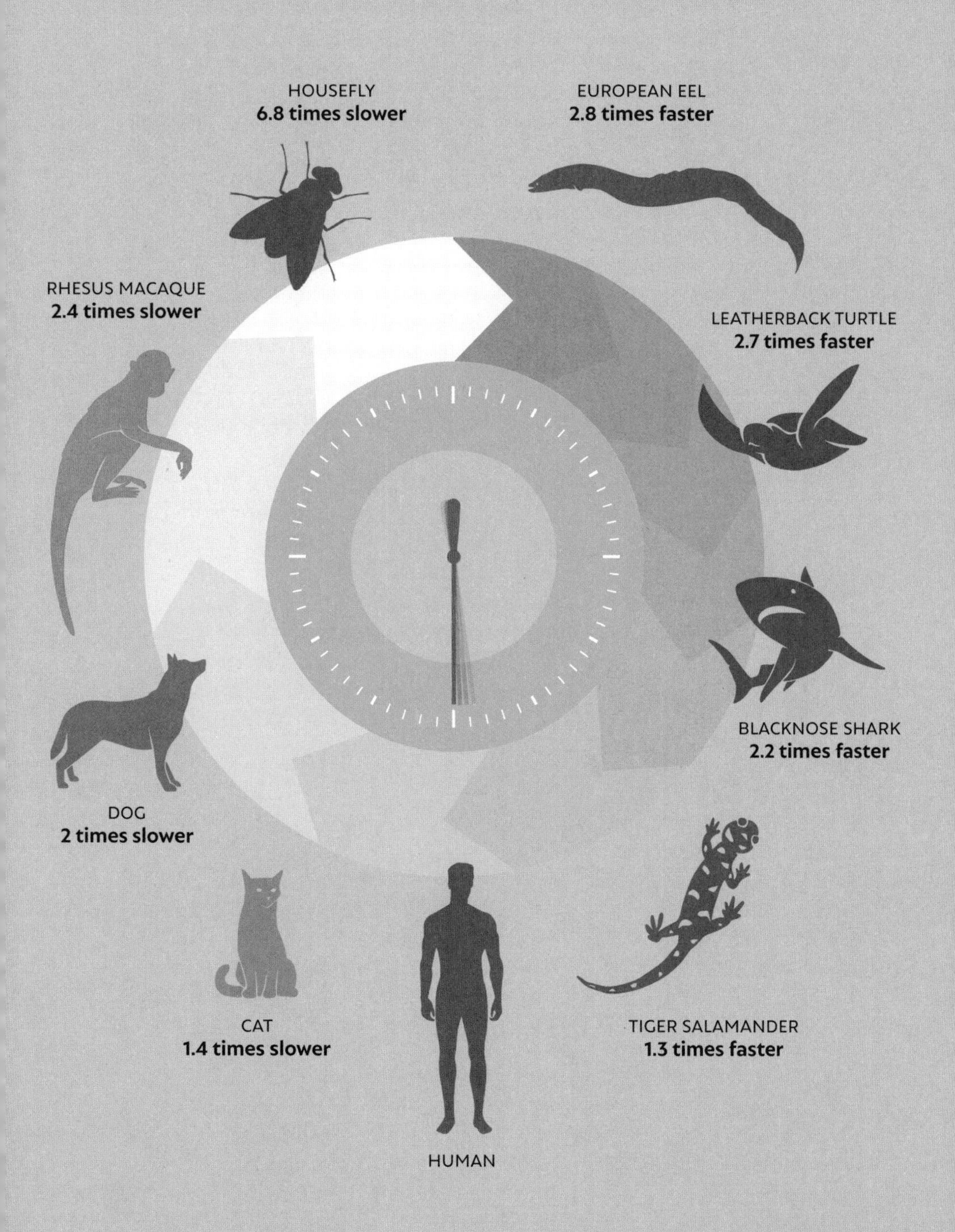

1
EVOLUTIONARY SPANS

INTRODUCTION

There has been life on planet Earth for more than 3.5 billion years, and during this time, our planet has been home to hundreds of millions of species. Today, it is estimated that about 9 million different species inhabit Earth, and that more than 98 percent of all species that have ever lived are now extinct.

The Earth has witnessed the emergence of life in its original single-celled form, then the explosion of diversity that occurred after single cells started collaborating, spawning the emergence of multicellular organisms. Our planet stood by as living things crept out of the oceans and started to thrive on land. It was then the ancestors of whales and dolphins left the land and returned to water. The Earth has seen the rise of plants, the demise of the dinosaurs, and the emergence of our own species. In its time, our planet has been home to fungi as tall as trees, beavers the size of bears, and dragonflies as big as barn owls.

So many fascinating life forms have evolved through evolution. Evolution is the process that enables living things to change over time. Sometimes, it occurs over unimaginably enormous time spans, with changes so subtle and slow, they can be hard to appreciate. Other times, it can be measured across much shorter time spans; sometimes evolution can be witnessed within a single human life span. As we will discover, life can evolve across millions and billions of years, but it can also change over decades and even within a single year.

The ancestor of whales was a land-dwelling mammal that returned to the sea around 50 million years ago.

LIFE IN A DAY

The Earth formed approximately 4.5 billion years ago, as material left from the formation of the Sun collided and fused together. In the time that followed, our planet has witnessed immense change, including the birth of our moon, the creation of our oceans, and, of course, the evolution of life. But if the history of the Earth was compressed into a single 12-hour time span, how would that 'day' pan out?

In the earliest hours, Earth takes a pummeling as volcanoes erupt and meteorites rain down. The Earth is hot and it takes time to cool. The first life appears around 2:45, in the form of single-celled organisms called "prokaryotes."

12 hours of evolutionary milestones

Two hours later, one prokaryote engulfs another, to create the very first cells containing a nucleus, called "eukaryotic cells." Many hours later, around 10:25, cells club together and start to collaborate. The first multicellular life forms emerge, paving the way for the full range of diversity that we see today. Our own species, *Homo sapiens*, is just a blip in an otherwise action-packed day. All of human history fits into just a few brief seconds, and a single human life is so fleeting that it barely registers at all.

2:45
Prokaryotes are the first life forms to emerge.

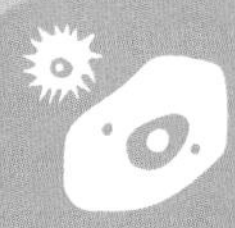

4:50
Eukaryotic cells, which contain a nucleus, evolve next.

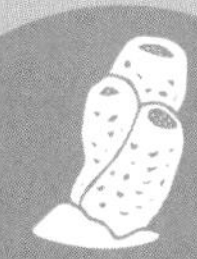

10:25
The evolution of multicellular life.

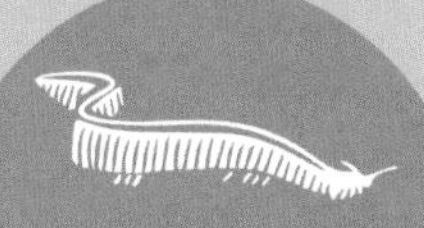

10:34
Life explodes into many different forms, such as arthropods and fish.

10:40
Some animals and plants move out of the ocean and onto land. The first insects appear.

11:00
The first four-limbed creatures evolve and head out of the water.

11:10
Reptiles evolve, paving the way for the rise of the dinosaurs.

11:32
The world welcomes the first mammals.

11:36
Birds evolve and take to the skies.

11:57 and 57 seconds
Anatomically modern humans evolve.

GEOLOGICAL TIME SPANS

The time spans we experience during our daily lives are neat and orderly. Each day comprises 24 hours, each hour contains 60 minutes, and every minute consists of 60 seconds. The Earth's history, however, is measured in geological time spans. These are big, unwieldy time frames running to millions and billions of years ("MYA" means millions of years ago; see below). To make life easier, scientists have divided these time periods into smaller, more manageable chunks.

Eons, such as the Phanerozoic, are the biggest time periods. They are subdivided into eras, which are in turn divided into periods and epochs. Unlike the orderly time slices we are used to, these geological time slices are not even in length. No two eras, for example, contain the same number of years, as geological time

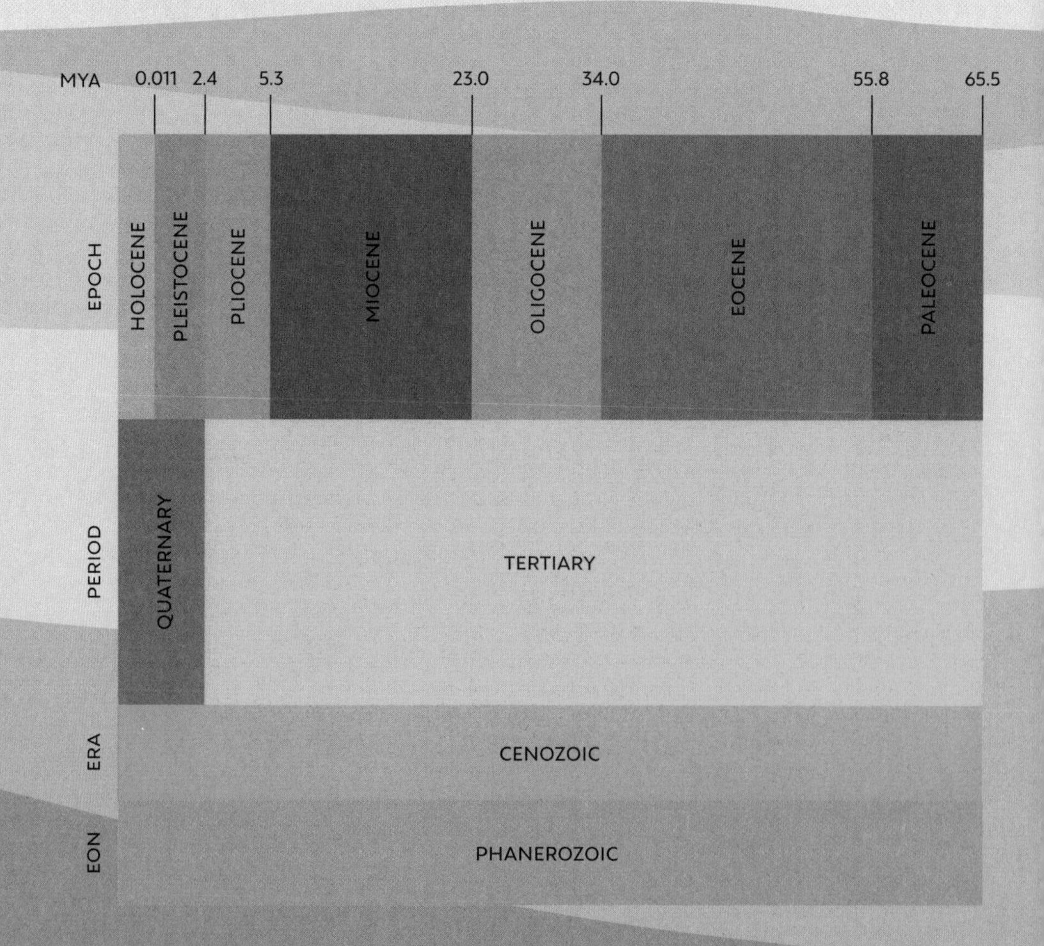

is divided using significant events from the Earth's history. For example, a massive burst in the diversity of life on Earth marked the beginning of the Paleozoic era.

Scientists have learned about the Earth's history by studying the layers of rock it is made from. Distinct layers were created at different times, so they contain a variety of fossils and have diverse chemical signatures. This helps scientists to age the rocks and trace the story of life on Earth.

We live in the Holocene epoch, but many scientists argue that human activities, such as the burning of fossil fuels and deforestation leading to climate change, have caused the Earth to shift. So, we are entering a new epoch, known as the Anthropocene. *Anthropo* comes from the Greek word for "human," and *cene* means "new."

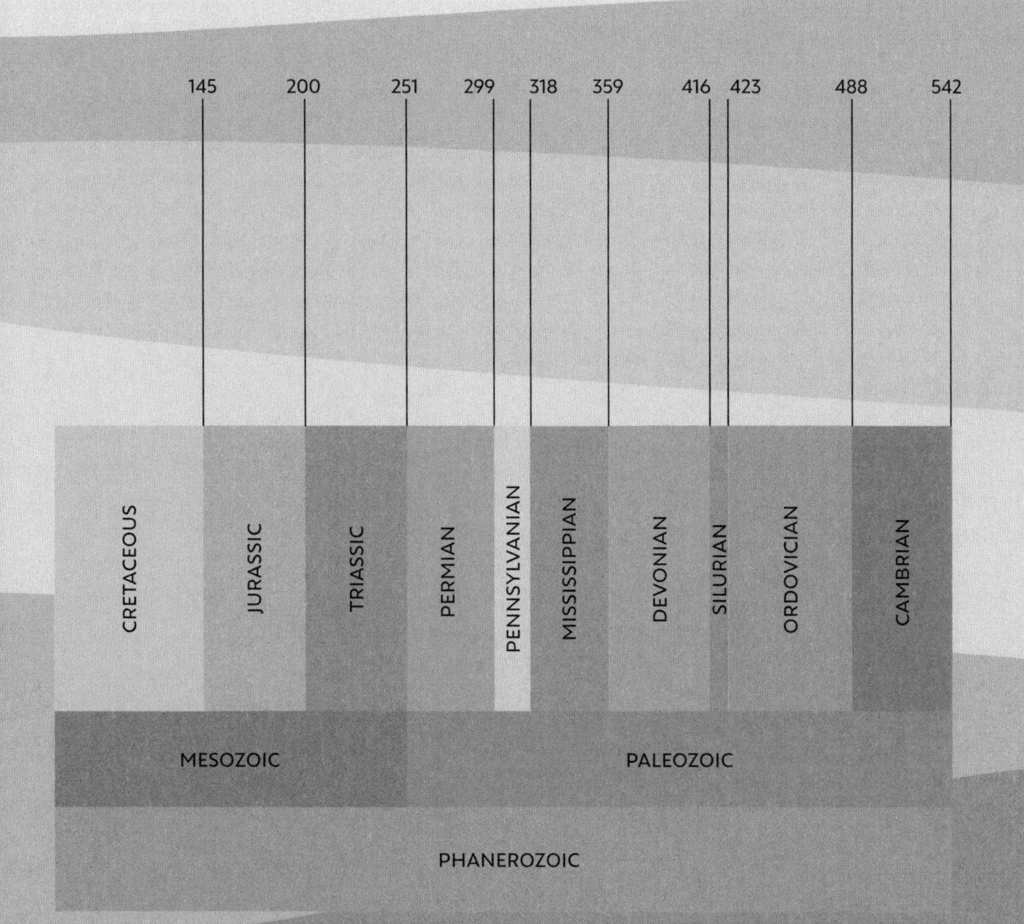

EVOLUTION IN ACTION

In 1859, British scientist Charles Darwin published his theory of evolution by natural selection. It explains how living things change over time, and how all the species that have ever existed came to be. It has become one of the most successful scientific theories.

Darwin noticed that although members of the same species are very similar, there are still differences. Some individuals, for example, may be bigger, faster, or better at finding food than others.

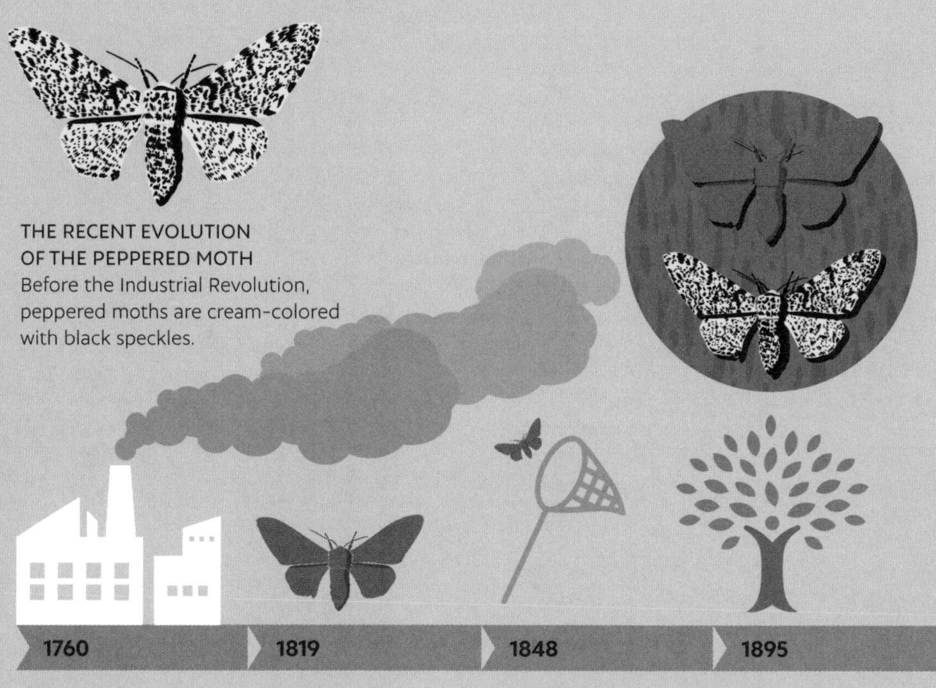

THE RECENT EVOLUTION OF THE PEPPERED MOTH
Before the Industrial Revolution, peppered moths are cream-colored with black speckles.

1760
The Industrial Revolution begins. Soot from chimneys settles on tree trunks where peppered moths rest during the day.

1819
A random mutation occurs in a gene linked to pigmentation. Moths with this DNA change have dark gray wings. This dark pigmentation is known as "melanism."

1848
A moth collector in Manchester spots the dark gray variation for the first time. This melanic form starts to become more common.

1895
Darker moths blend in with the dirty tree trunks. They are less likely to be eaten and more likely to survive and reproduce. In Manchester, 98 percent of peppered moths are melanic.

Darwin argued that these same individuals are more likely to survive, reproduce, and pass on their useful characteristics to future generations, compared to individuals that are less well-adapted to their environment.

This theory is known as natural selection. Over time, features that boost survival become more common, and species gradually change. Eventually it can lead to the evolution of new species. The peppered moth is a great example of evolution in action.

The Clean Air Act is introduced. Pollution levels drop. The tree trunks become cleaner and lighter.

Darker moths are more obvious and likely to get eaten. Lighter-colored moths are more likely to survive and reproduce, so now they become more common over time.

Although there is some variation, now peppered moths are mainly cream-colored with tiny, peppery black spots.

THE EVOLUTION OF THE HORSE

The peppered moth evolved very quickly in response to the rapid environmental changes that occurred during the Industrial Revolution. This is called "contemporary evolution" because it can be seen across the span of a human life.

Darwin would have been amazed. He thought that evolution happens slowly, across many millions of years. He once said, "We see nothing of these slow changes in progress, until the

EVOLUTION: 50 MYA
NAME: *Hyracotherium*
HEIGHT: 1.3 feet (0.4 m)
NUMBER OF FOREFOOT TOES: 4
CHEEK TEETH: Small

EVOLUTION: 35 MYA
NAME: *Mesohippus*
HEIGHT: 2 feet (0.6 m)
NUMBER OF FOREFOOT TOES: 3
CHEEK TEETH: Small

EVOLUTION: 15 MYA
NAME: *Merychippus*
HEIGHT: 3.3 feet (1 m)
NUMBER OF FOREFOOT TOES: 3
CHEEK TEETH: Medium-size

Body size increases and limbs lengthen, as primeval forests give way to vast grassland areas and the equines need to outpace predators.

Early horses walked on several spread-out toes—an adaptation to moving around on the soft, wet forest floors. Over time, toe number decreases. Single-toed limbs make it easier to trot while roaming around, looking for food and water.

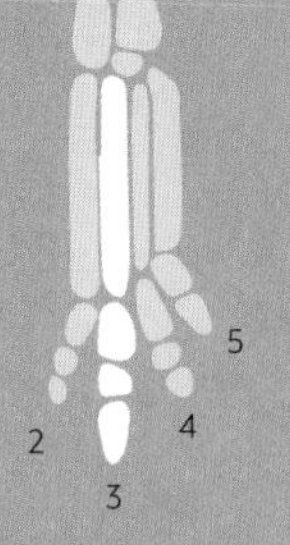

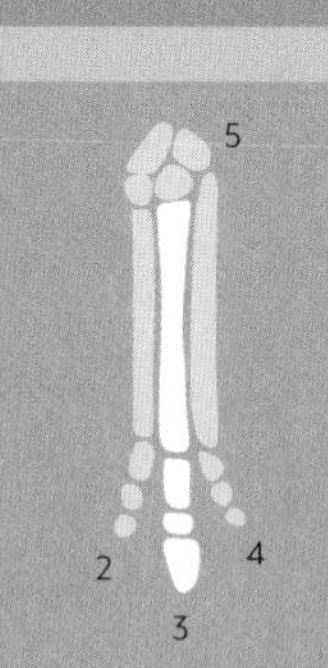

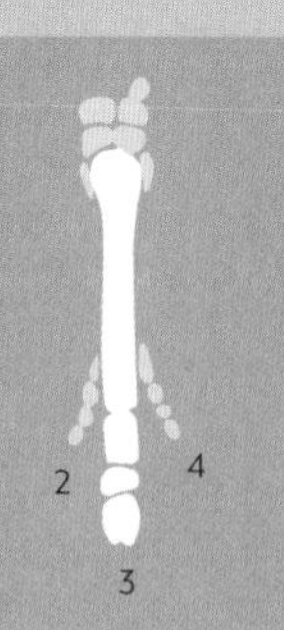

Teeth become bigger and more durable as early horses transition from eating soft leaves to tougher, more abrasive grasses.

hand of time has marked the long lapse of ages." The only way to "see" evolution, Darwin thought, was by studying fossils, which are laid down across vast, geological time frames.

The evolution of the horse is a great example, because it is well documented in the fossil record. The story starts 50 million years ago, with a small dog-size, forest-dwelling equine, which went on to evolve into the large, majestic animals we have today.

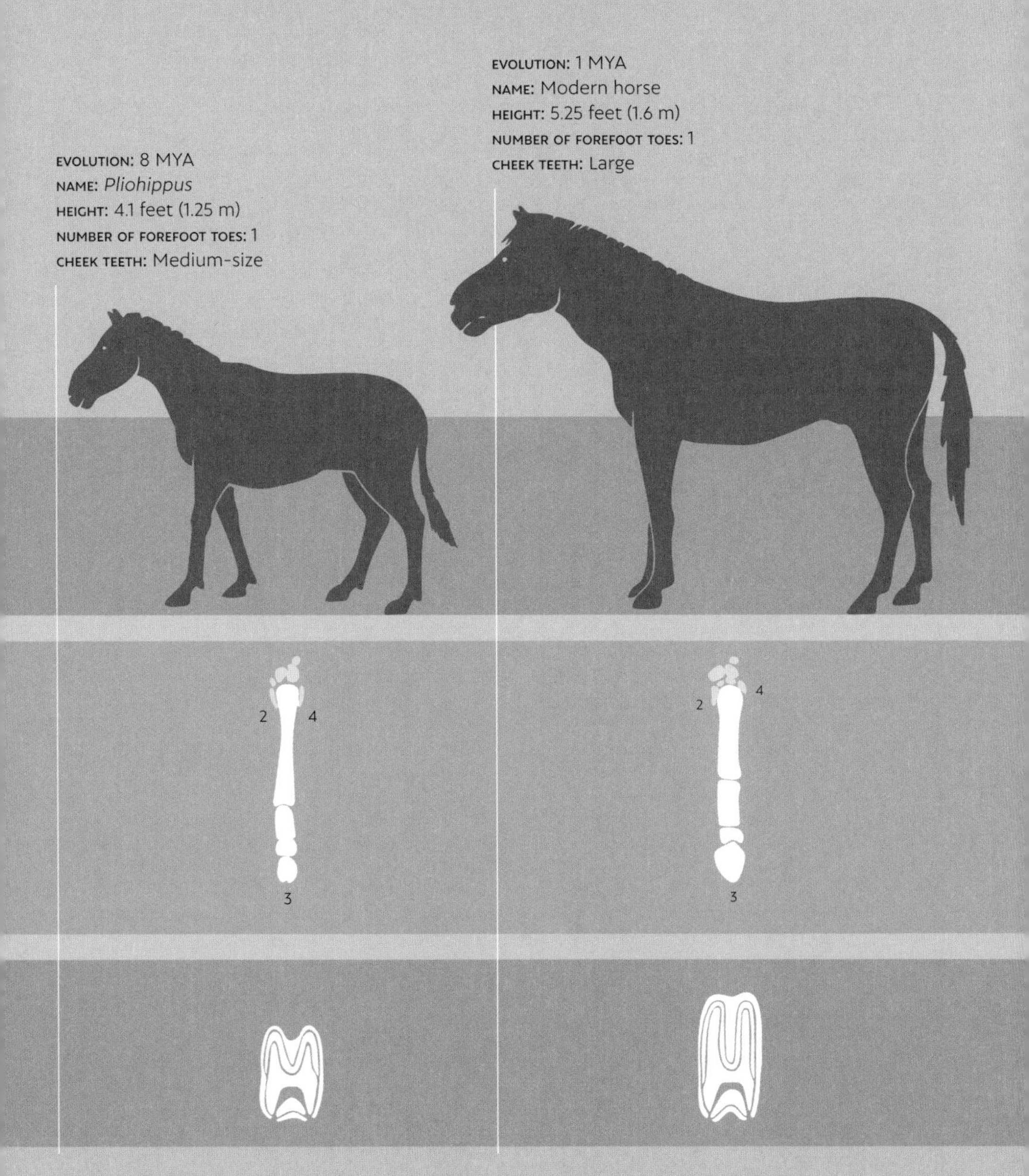

MASS EXTINCTIONS

Extinction is a normal part of life. About 98 percent of all the species that have ever lived on our planet have died out. New species evolve all the time, and older ones become extinct when they are outcompeted or unable to adapt to change. There is a normal background rate of extinction, thought to be between 0.1 and 1 per 10,000 species every 100 years.

Sometimes, however, extinction rates spike, and species die out more quickly than they are replaced. These are known as periods of mass extinction, occurring when more than 75 percent of the world's species become extinct in a "short" amount of geological time—less than 3 million years. Recovery is slow but

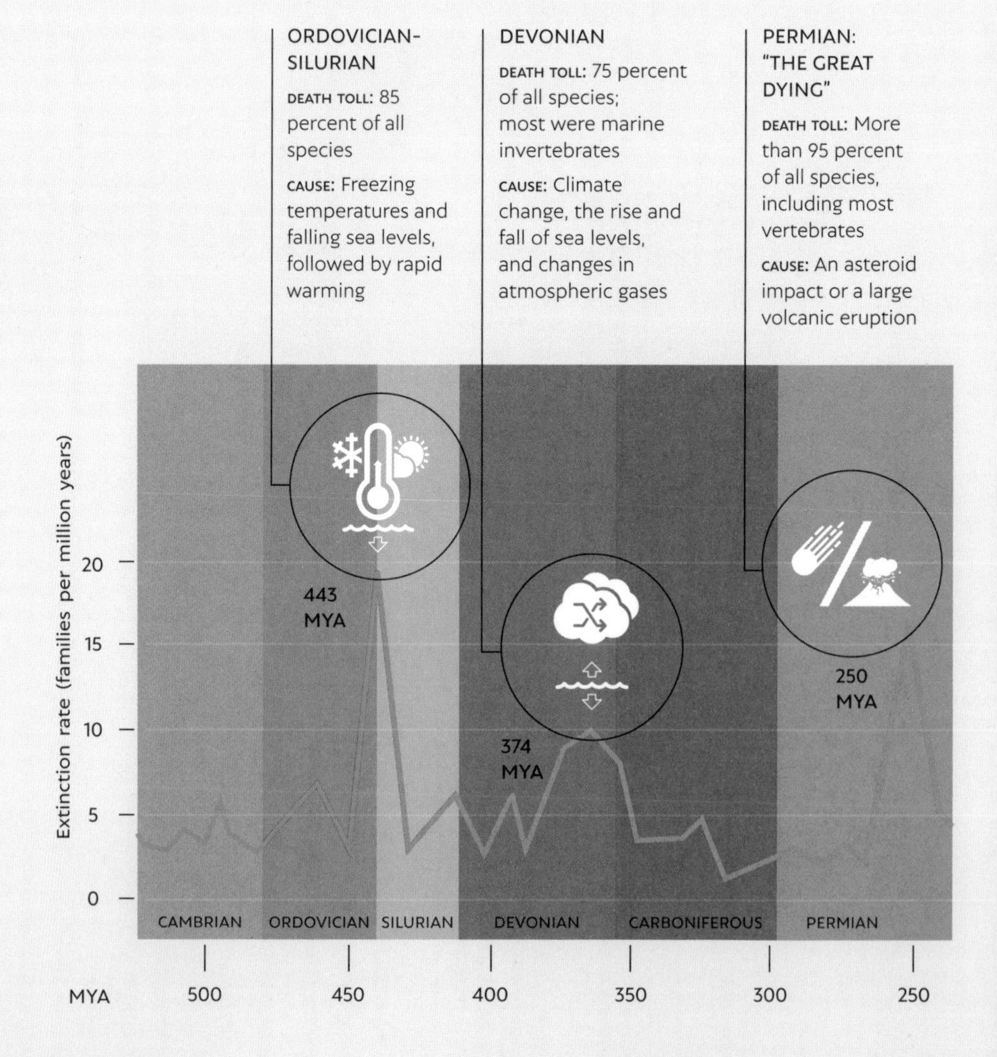

possible. It is believed that ecosystems take around 2 million years to become functional and resilient after a mass extinction event.

There have been five mass extinctions, all caused by natural events, but scientists are concerned that we are living through a sixth mass extinction that is entirely of our own making. Human activities, such as the burning of fossil fuels, that result in climate change—and the unsustainable use of energy, water, and land—are causing extinction rates to soar. Rates are now up to 1,000 times higher than during prehuman times, and it is estimated that we lose between 30 and 150 species every day. We are destroying the very ecosystems that we depend on for our survival.

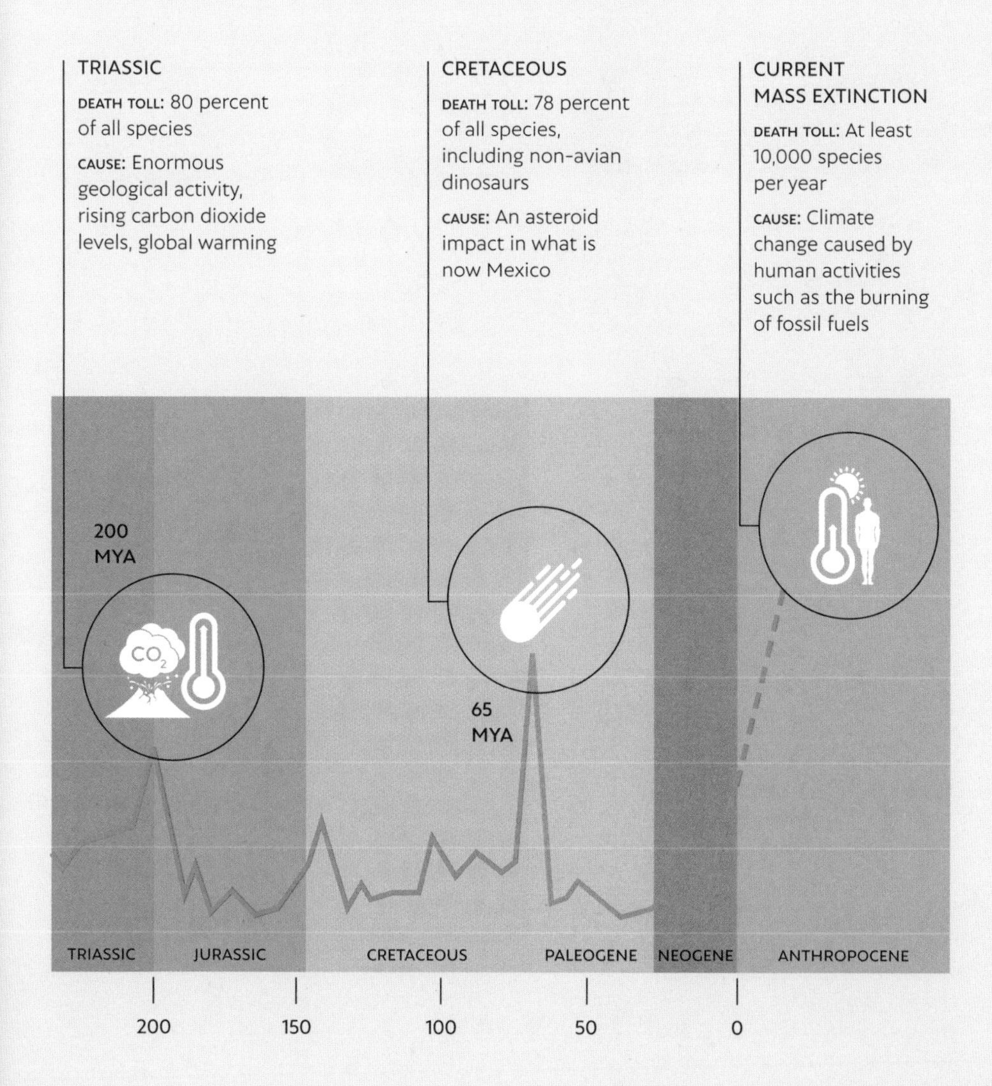

THE RISE AND RISE OF PLANTS

The biggest bloom in the world belongs to the corpse flower, *Rafflesia arnoldii*: a rare plant found in the rain forests of Indonesia. It smells of decaying flesh, grows up to 3 feet (1 m) wide, and belongs to the most diverse group of land plants that exists today. Flowering plants, or angiosperms, are one of evolution's biggest success stories. Appearing only in the last 140 million years, they are youngsters in the bigger story of life on Earth, yet in that time, their group has diversified to produce more than 300,000 different species.

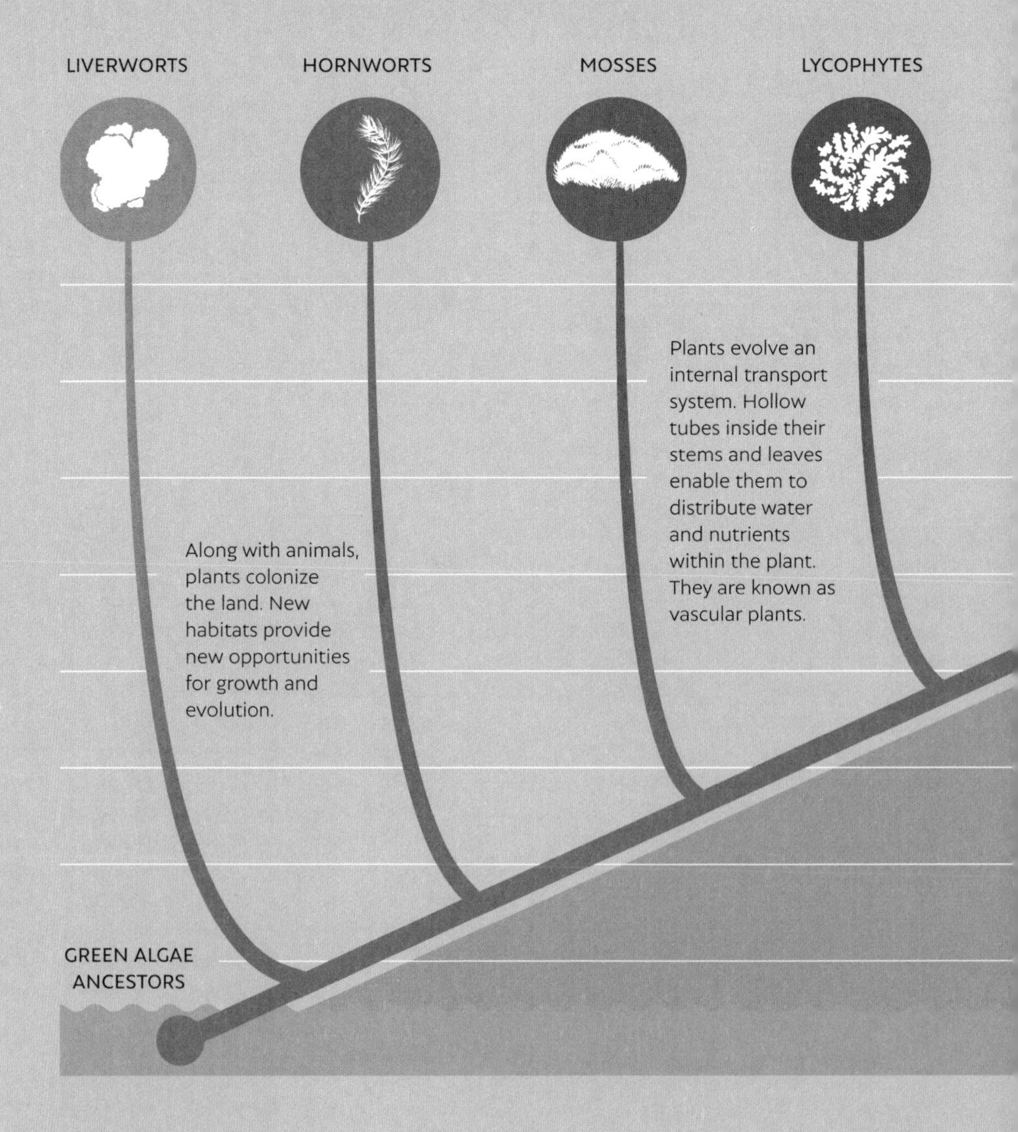

All plants evolved from common, green algae ancestors that lived in oceans over 500 million years ago. Life moved onto land, opening up new opportunities. The earliest land-living plants were small and filamentous. They had no flowers, roots, or way of moving water around internally, so their size was limited and they were tied to damp environments, just like modern liverworts. After plants evolved internal transport systems and seeds around 300 million years ago, it became easier to grow larger and colonize new habitats. Now there are flowers that grow in the desert, and the world's biggest tree—a giant sequoia called General Sherman—is over 263 feet (80 m) tall.

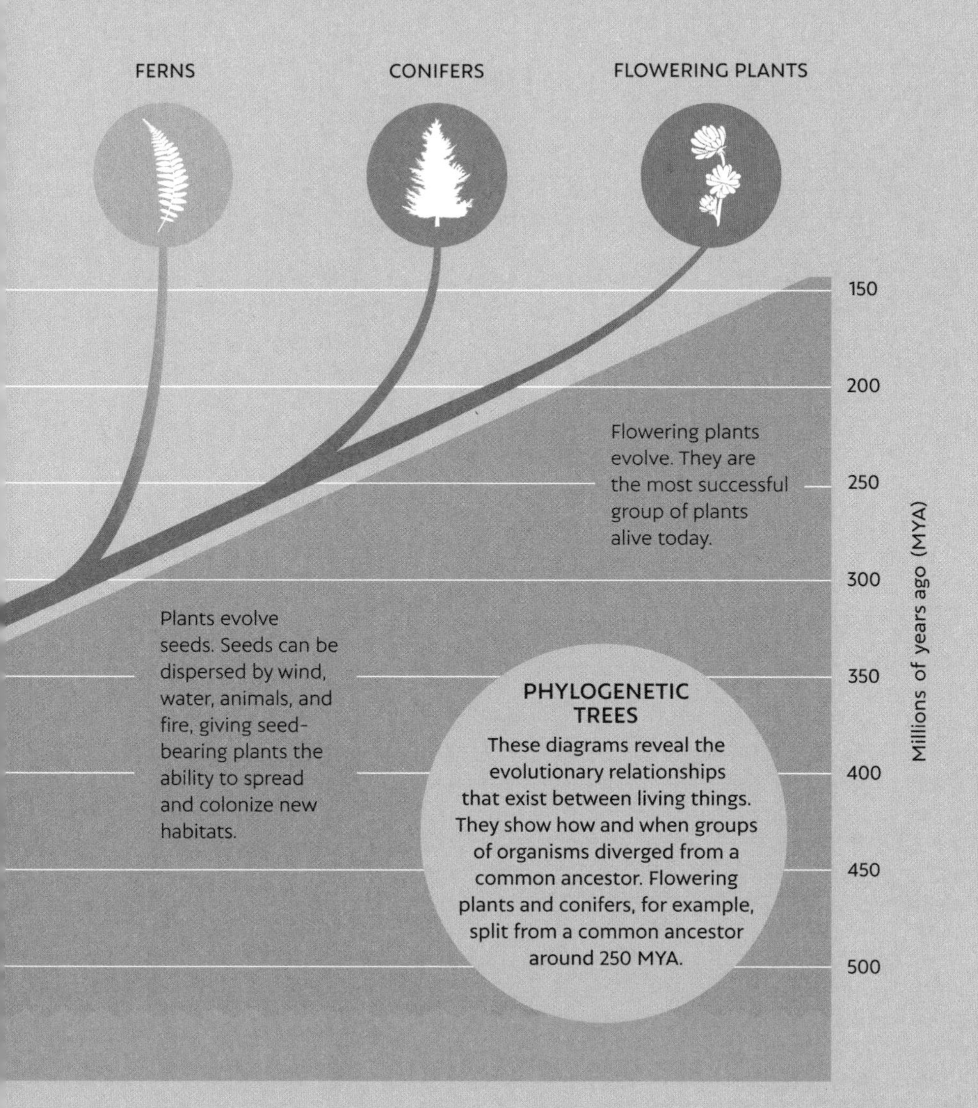

TERRIBLE LIZARDS

Dominating the planet for over 170 million years, dinosaurs were one of the most successful groups of land animals ever. They lived during the Mesozoic era, which stretched from around 251 to 65 million years ago. The Mesozoic era is divided into three periods: Triassic, Jurassic, and Cretaceous. These periods are often split into Early, Middle, and Late subdivisions, and these subdivisions are divided into smaller slices of time called "stages." Each stage lasts around 5 million years, and specific dinosaur species are unique to certain stages. *Tyrannosaurus* and *Triceratops*, for example, belong to the final stage of the Late Cretaceous, called the Maastrichtian.

Dinosaurs in the Mesozoic era

251–201 MYA

TRIASSIC PERIOD

The world's continents are united as a single supercontinent called Pangea, surrounded by a vast ocean. The climate is relatively hot and dry, and much of the land is covered by desert. Dinosaurs evolve from reptiles. At 243-million-years-old, *Nyasasaurus parringtoni* is thought to be either one of the earliest dinosaurs or closely related to them. It was a leggy, long-necked creature that measured 6–10 feet (2–3 m) from tip to tail. Other dinosaurs from this period include *Coelophysis*, *Eoraptor*, and *Herrerasaurus*.

NYASASAURUS PARRINGTONI

DINOSAURIA

In 1841, Victorian biologist and fossil fan Richard Owen coined the term "dinosauria" meaning "terrible lizard."

201–145 MYA

JURASSIC PERIOD

Pangea starts to break apart, becoming a northern continent called Laurasia and a southern continent called Gondwana. Dinosaurs start to evolve into northern and southern variants. The climate is warm and tropical, giving rise to lush vegetation. Dinosaurs rise to dominate the land and some grow to huge sizes. *Apatosaurus*, for example, was a long-necked, plant-eating sauropod dinosaur that lived in the Late Jurassic. It grew up to 75 feet (23 m) long. Other iconic dinosaurs from this period include *Brachiosaurus*, *Diplodocus*, and *Stegosaurus*.

145–66 MYA

CRETACEOUS PERIOD

During this time, the continents continue to break and drift apart. Some of today's continents are recognizable, but they lie in different positions. Dinosaurs continue to rule the land and evolve into further forms. Famous Cretaceous dinosaurs include *T. rex*, *Triceratops*, *Spinosaurus*, *Ankylosaurus*, and *Velociraptor*. It doesn't end well for them though; after a 7-mile-wide (12 km) asteroid pummels the Earth, all non-avian dinosaurs become extinct.

DINOSAUR GROWTH

Evolution favors any strategy that helps a species to survive, reproduce, and pass on its genes to future generations. The tyrannosaurs, including *T. rex*, evolved an unusual strategy to do this. Recent research suggests that juvenile and adult tyrannosaurs looked and behaved quite differently to one another, hunting down and eating different types of prey. This enabled them to live side by side, in the same environment, with minimal competition between the generations. A single species of tyrannosaur occupied the space that would typically be claimed by two different species of predator. This created ecosystems quite unlike anything seen today.

NEW SPECIES

Around 1,500 species of dinosaurs have been described so far, but this is just the tip of the iceberg. Most animals die without becoming fossilized, and many of the fossils that do exist are still waiting to be found. New fossils are being unearthed all the time, and on average, scientists describe one new species of dinosaur per week.

Skull shape across the years

LIFE STAGE: Juvenile
AGE SPAN: Up to 11 years
BITE FORCE: Relatively weak; 5–13 percent of adult value
TEETH: Thin, bladelike, and brittle
BEHAVIOR: Juveniles are small and agile. They hunt relatively small prey.

Jaw length

20–60 CM

LIFE STAGE: Teenage
AGE SPAN: 11–19 years
BITE FORCE: Intermediate; 13–100 percent of adult value
TEETH: Thicker and more robust
BEHAVIOR: Teenagers grow rapidly and become able to hunt larger prey.

60–85 CM

DIETARY SHIFT

Juveniles eat small prey such as tiny dinosaurs, amphibians, and lizards.

A CASE OF MISTAKEN IDENTITY?

T. rex may well be one of the most iconic and best-known dinosaur species, but it's in the midst of an identity crisis. A study published in 2022 suggested that the Cretaceous heavyweight, famed for its mini arms, might not be one but three separate species—a redefined *T. rex*, *T. regina*, and *T. imperator*. The theory, which is based on analyses of fossil leg bones and teeth, predicts that *T. rex* and *T. regina* lived side by side in the Late Cretaceous and evolved from *T. imperator*. Many dinosaur experts, however, are not convinced, so further research is needed.

LIFE STAGE: Adult
AGE SPAN: 20–35 years
BITE FORCE: Strong enough to crush bone; 100 percent of adult value
TEETH: Not to be messed with
BEHAVIOR: Adult animals have deep skulls and powerful bites. They can take down really large, herbivorous dinosaurs such as *Triceratops*.

Adults eat large herbivorous dinosaurs such as *Triceratops*.

TIME FOR TAKEOFF

Flight is so useful to animals that it has evolved on at least four separate occasions: first in insects, then pterosaurs, followed by birds and then bats. The development of flight is a story that spans at least 400 million years, and even today, flight continues to evolve in those species that have taken to the skies.

The evolution of flight

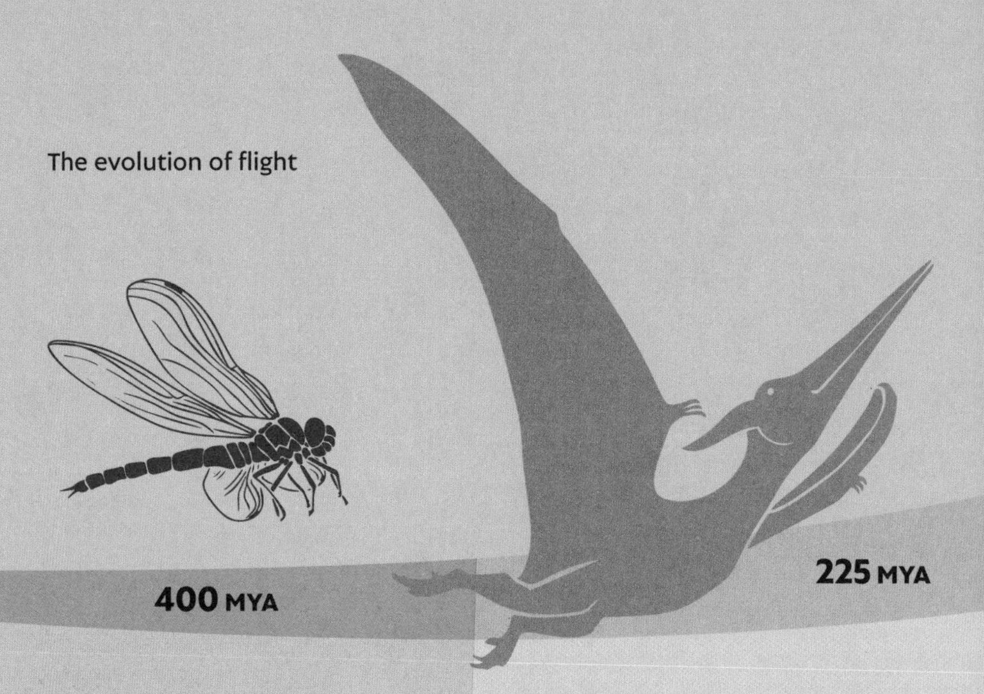

INSECTS

Flying insects evolved during the Carboniferous period. No one knows exactly how insects evolved the ability to fly. One idea is that the precursors to wings evolved in insects that lived on the water's surface, where they may have "caught the wind," which helped to propel the insect along. Another is that they evolved in tree-dwelling insects, and they may have helped the insect to parachute down to the ground. Today, there are an estimated 10 million insect species.

PTEROSAURS

Pterosaurs were close cousins of the dinosaurs. Both groups of reptiles lived around the same time, but only pterosaurs took to the skies. They were the first vertebrates to do so, evolving during the Triassic period. Their group lasted for over 140 million years, making them incredibly successful, and some grew to enormous sizes. *Quetzalcoatlus northropi*, for example, is thought to be the largest flying animal ever. With its 36-foot-long (11 m) wingspan, it had to jump up to 8 feet (2.5 m) in the air in order to take off.

50 MYA

165–150 MYA

BIRDS

Birds evolved during the Jurassic period, from a diverse group of dinosaurs known as "therapods." As a result, the first birds shared many features with dinosaurs, such as sharp teeth and long bony tails. Their classic lightweight, feathered, and winged body plan took shape gradually over the next 10 million years. They then diversified into many varieties, and today there are more than 10,000 different species of birds.

BATS

Bats are the only mammals to truly fly. Unlike birds, their evolution is not well documented in the fossil record, but it's thought that bat-like mammals were on the wing during the Eocene epoch. Scientists think they may have acquired the ability from a tree-living ancestral species that used winglike structures to carefully flutter to the ground in order to forage. Today, there are more than 900 species of bats.

THE MECHANICS OF FLIGHT

Insects, pterosaurs, birds, and bats didn't evolve from the same single flying ancestor. Instead, over many millions of years, they all evolved the ability to fly from different ancestors that were unable to fly. It's a classic example of convergent evolution, where various species independently evolve similar features via different means. Each group created independent solutions to the problem of being unable to fly.

Insects evolved one successful strategy with their four separate wings. Birds, bats, and pterosaurs, on the other hand, all have two wings that are modified forelimbs, but they are all modified in different ways.

FLYING

Once animals were able to fly, it proved tremendously useful, allowing them to escape predators, for example. This is thought to be a major driving factor in the evolution of bird and insect flight. Meanwhile, the ability to catch food on the wing is thought to have spurred the evolution of flight in pterosaurs and bats.

Four solutions to the problem of flight

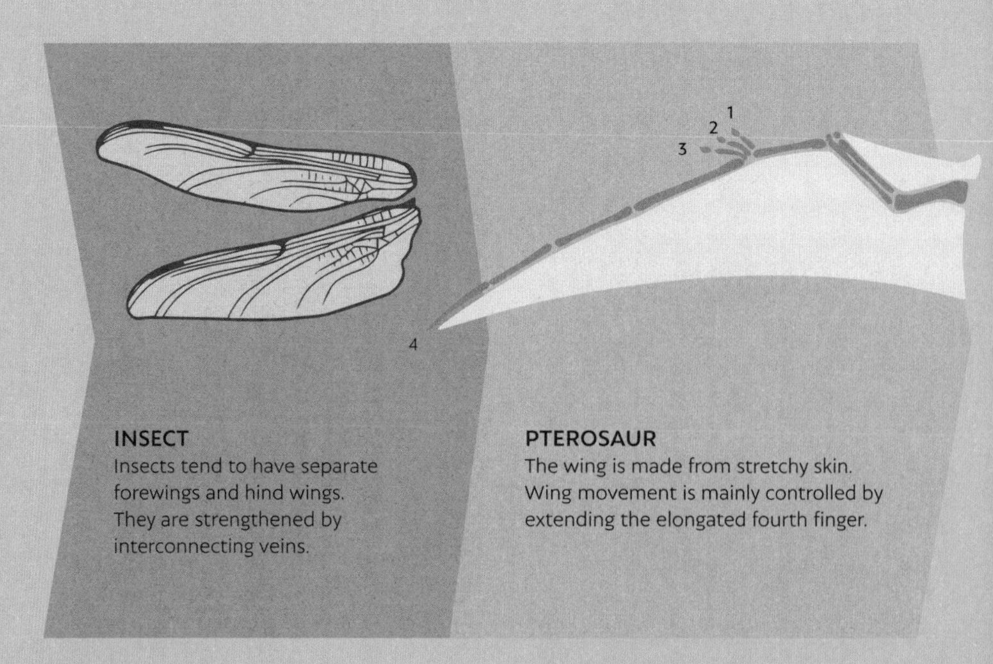

CLIFF SWALLOWS EVOLVING

Evolution never stands still, so flight is still evolving today. In Nebraska, cliff swallows build their little nests on the undersides of bridges and overpasses. Over a period of just a few decades, their wings have evolved and are now several millimeters shorter than they were 30 years earlier. As a result, the number of swallows killed each year on the road has declined, even though traffic volumes have increased. This is another example of contemporary evolution, where evolution occurs rapidly.

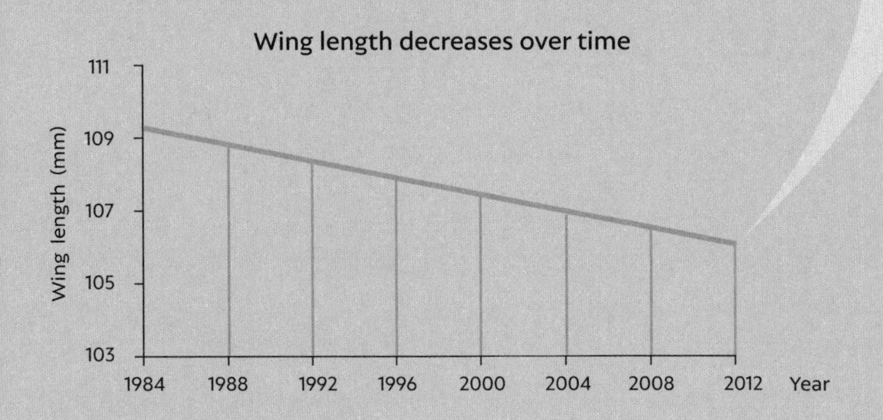

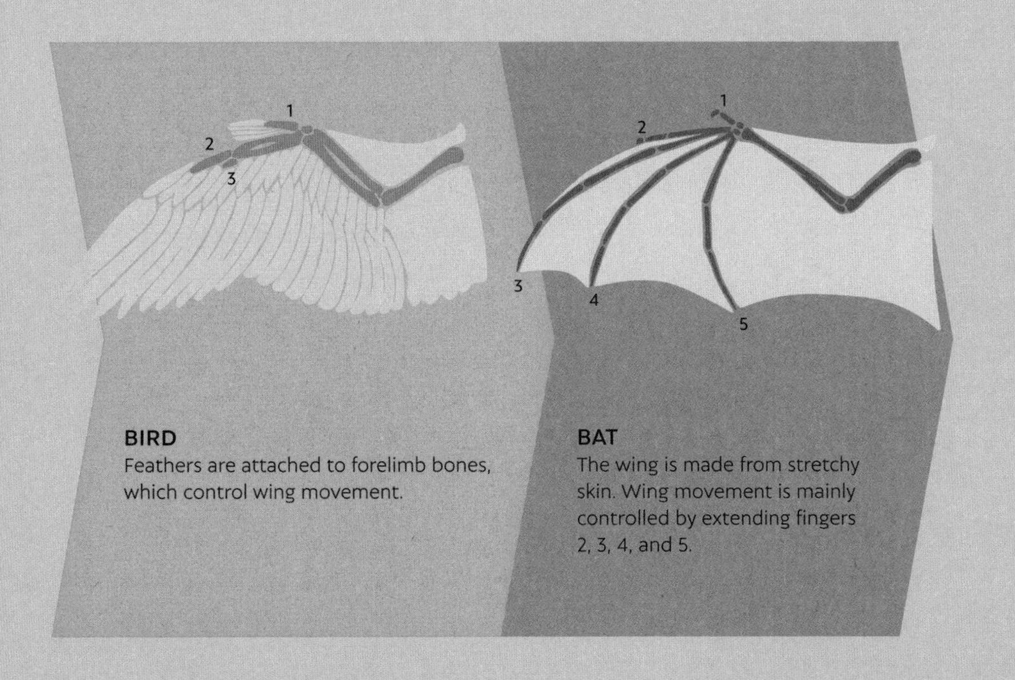

BIRD
Feathers are attached to forelimb bones, which control wing movement.

BAT
The wing is made from stretchy skin. Wing movement is mainly controlled by extending fingers 2, 3, 4, and 5.

ROCK STARS

Paleontology is the study of the history of life on Earth through fossils. The practice began more than 2,500 years ago, when ancient scholars, such as Xenophanes, wrote about the fossils of marine organisms, hinting that dry land was once underwater. The discipline became more scientific during the nineteenth century, and today, as new fossils and methods to study them continue to emerge, we still have much to learn about life's prehistoric past.

A timeline of fossil discovery

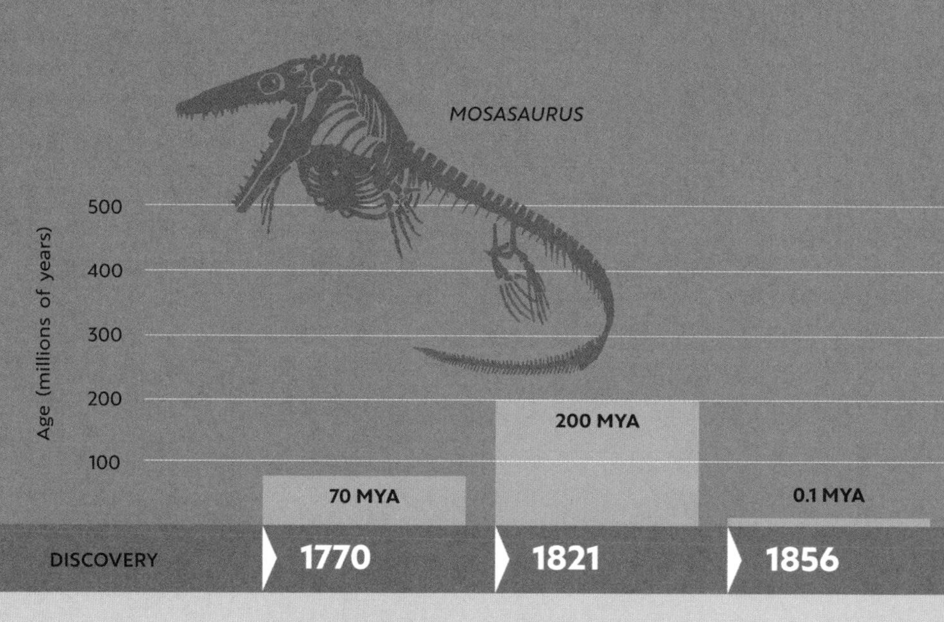

Mosasaurus (above) was a huge marine reptile. Its fossilized bones were found in a Dutch chalk quarry. They were initially thought to be crocodile or whale bones, but 50 years later, scientists realized the bones belonged to a marine reptile.

Plesiosaur was a long-necked marine reptile with four flippers and a tail. Victorian fossil hunter Mary Anning discovered the first skeleton in Lyme Regis, England. The fossil record now has more than 100 different plesiosaur species.

Homo neanderthalensis fossils were found in Germany's Neander Valley and named after it. Commonly called Neanderthals, they were an ancient type of human and are one of our closest relatives.

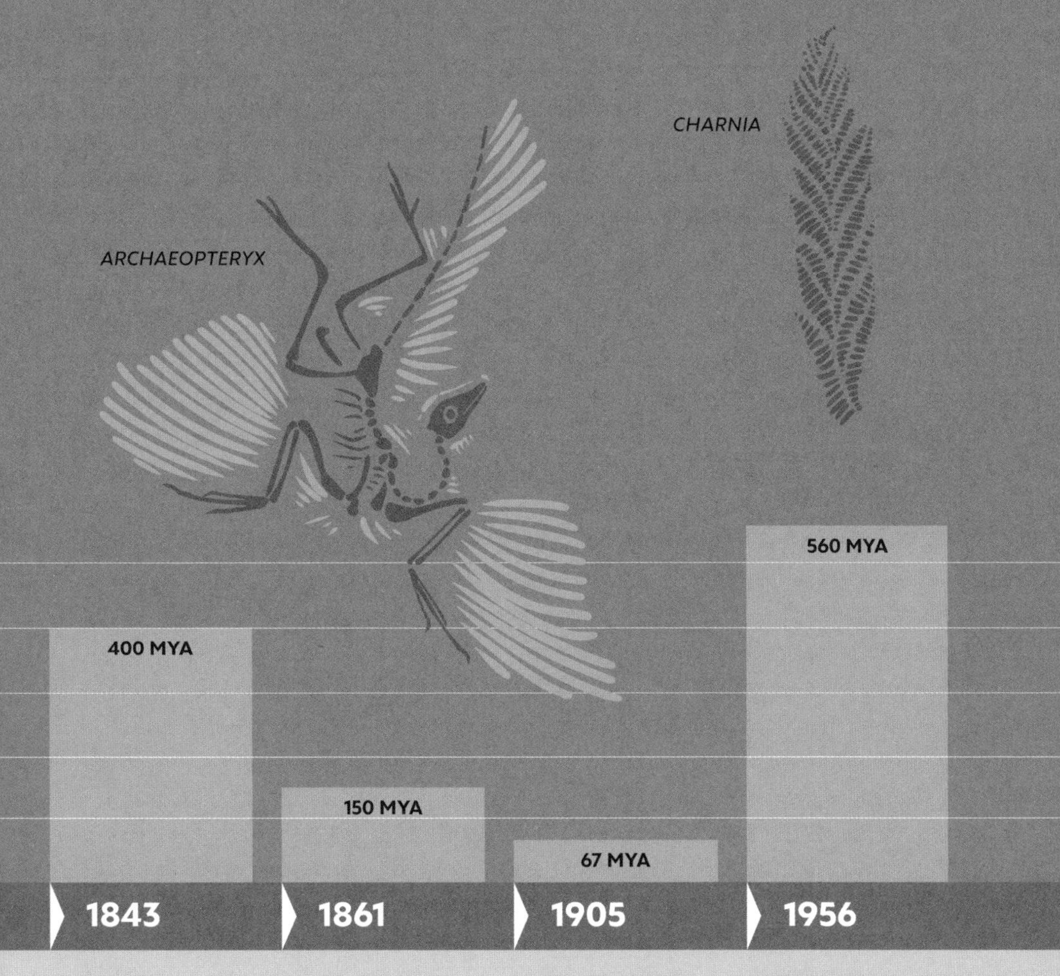

1843

Prototaxites was discovered in Canada. It grew up to 26 feet (8 m) tall, and this trunk-like life form towered over all other land-living organisms during its heyday. Scientists think it was either a humongous fungus or a lichen.

1861

Archaeopteryx (above) was the first skeleton and found in Bavaria, Germany. It displayed a mixture of dinosaur- and birdlike features, including teeth, feathers, and a bony tail. It is an important evolutionary link between non-avian dinosaurs and birds.

1905

Tyrannosaurus rex was described and named by Henry Fairfield Osborn. *Tyrannos* means "tyrant," *sauros* means "lizard," and *rex* means "king." With its tiny arms, *T. rex* became the world's most famous dinosaur.

1956

Charnia (above), a frond-shaped fossil, was discovered by Tina Negus in Charnwood Forest, UK. It is part of the Ediacaran biota, a bizarre group of organisms that are some of the earliest examples of complex multicellular life.

A timeline of fossil discovery

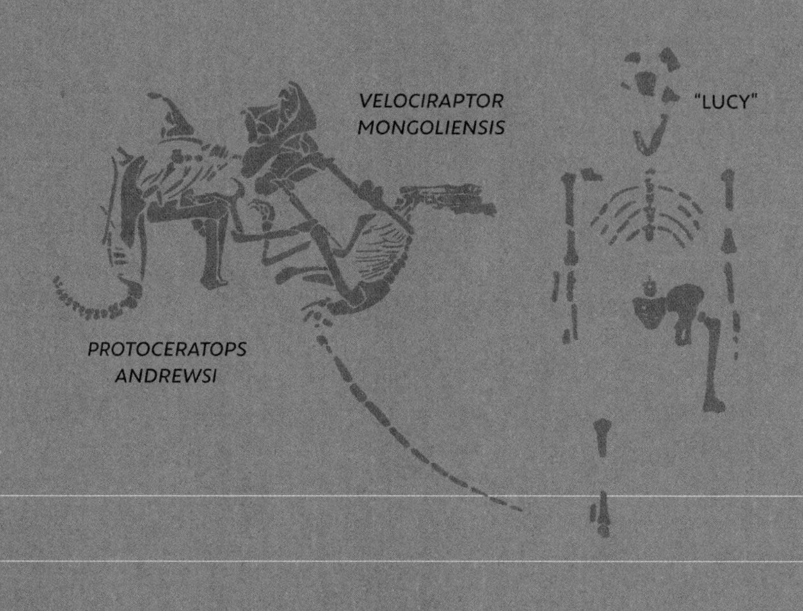

DISCOVERY	1971	1974	2003
	80 MYA	3.2 MYA	0.7 MYA
	Protoceratops andrewsi and ***Velociraptor mongoliensis*** (above) were unearthed in the Gobi desert. The remains of these two dinosaurs were locked in combat, providing direct evidence that velociraptors preyed on these large, herbivorous dinosaurs.	**"Lucy"** (above), a member of *Australopithecus afarensis*, was found in Hadar, Ethiopia. Hundreds of bones belonging to this humanlike female were discovered. She was the first of our ancestors to stand tall and walk out of the forest onto the savanna.	A fragment of horse leg bone jutting out of the permafrost in Canada's Yukon Territory was spotted by Eske Willerslev. Remarkably, he was able to retrieve DNA from the specimen. It is some of the most ancient DNA ever studied.

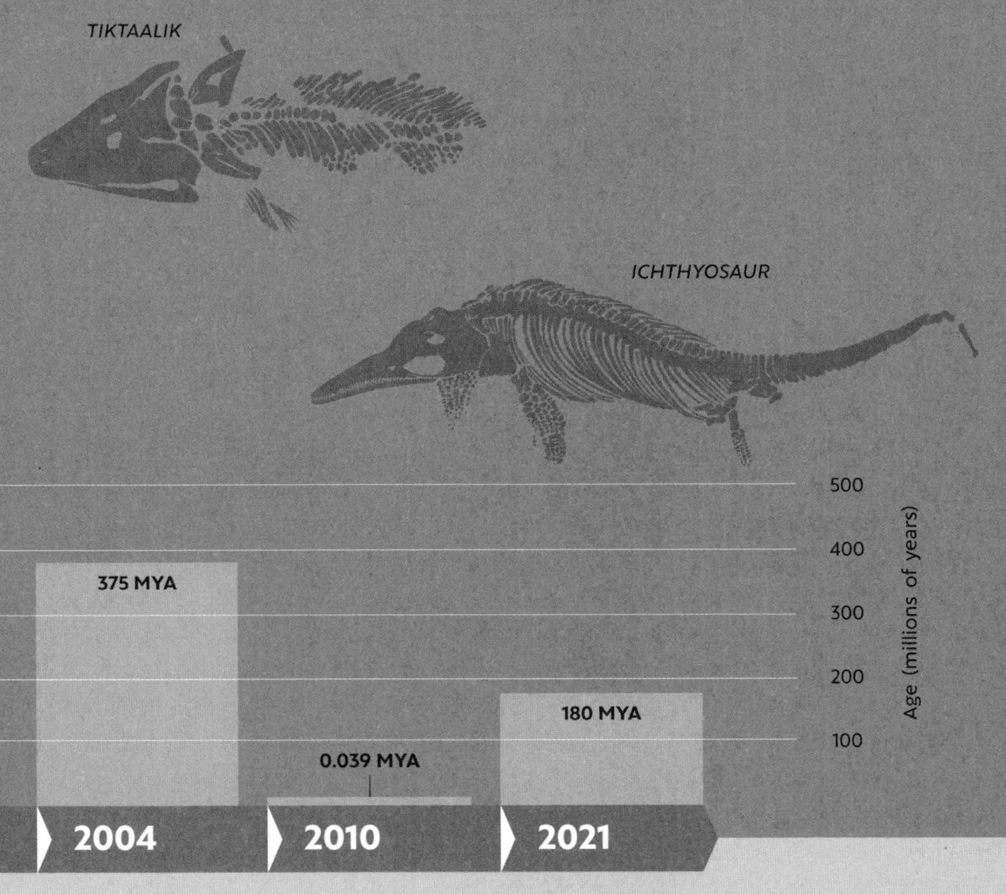

Tiktaalik (above) was discovered. It had gills, scales, primitive lungs, basic wrist bones for pushing its body up, and a sturdy ribcage for lumbering onto the shore. It is a transitional species between fish and the four-legged vertebrates that evolved from them.

Yuka is the best-preserved woolly mammoth carcass. It was discovered by Siberian tusk hunters. Her mummified remains contain skin, muscle, and fur. Analysis of her teeth shows that she was about 8 years old when she died.

Ichthyosaur (above), the largest, most complete marine reptile skeleton ever found in Britain, was discovered by Joe Davis. It was a 33-foot-long (10 m) sea predator and its skull alone weighs more than 1 ton.

FOSSILS

Fossils are not always made of stone. They are the preserved remains, impression, or trace of any previously-living thing from a past geological age. Mummified items are fossils, for example, as are some objects that have been preserved in amber or tar.

MONOTREMES, MARSUPIALS, AND MAMMALS

We are all familiar with mammals—the furry, milk-guzzling, warm-blooded group of beasts, of which we are a member. Yet some species are odder than others. Marsupials, such as wallabies and opossums, give birth to relatively undeveloped young that mature inside the mother's pouch, while monotremes, such as platypus and echidnas, lay eggs instead of giving birth to live young. Go back far enough in time, and they're all related—part of an increasingly divergent evolutionary journey that spans over 200 million years.

Evolution of monotremes, marsupials, and mammals

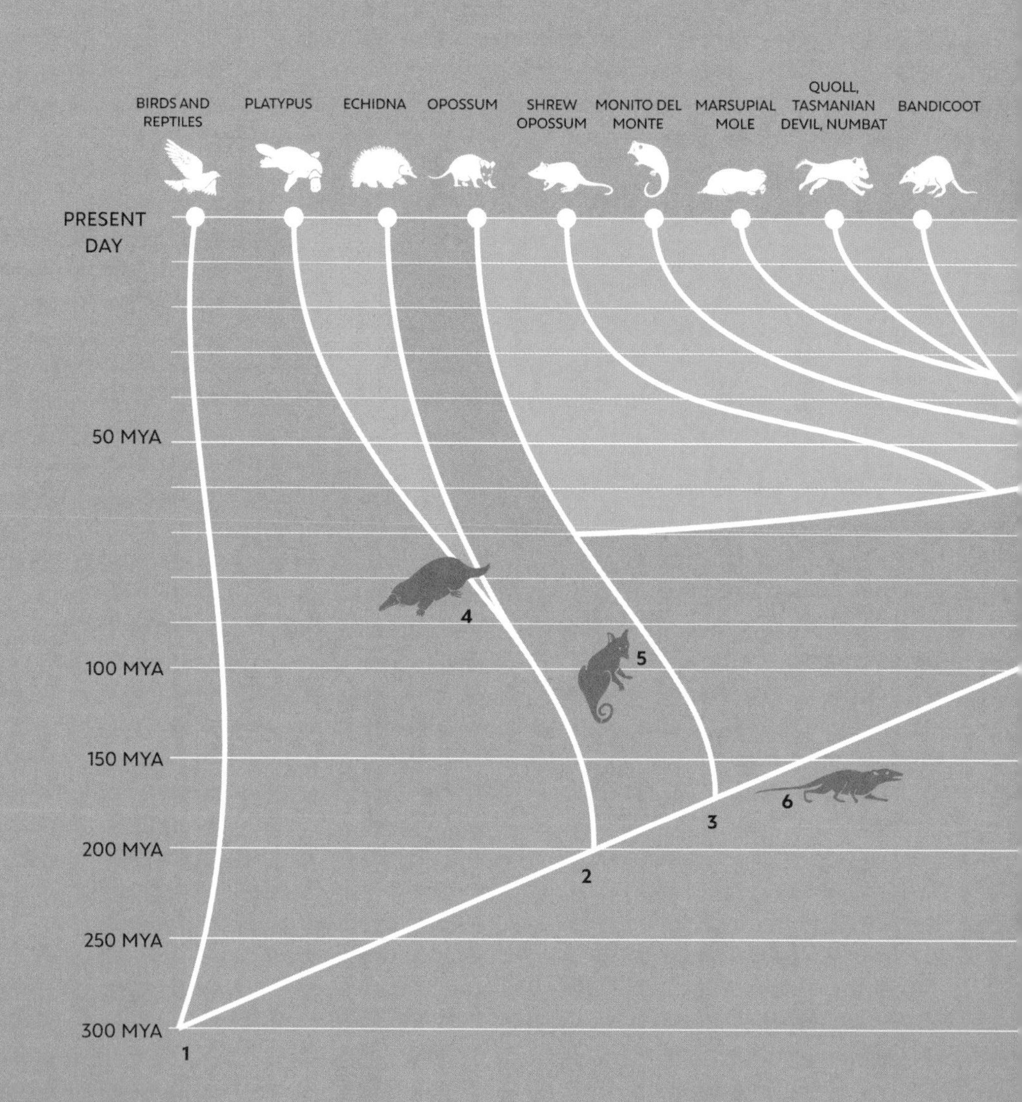

Marsupials evolved in North America around 170 millions years ago. Approximately 66 million years ago, they made their way to South America, which was then connected to Australia. Ten million years later, marsupials had arrived in Australia. The oldest Australian fossil marsupials, which are 55 million years old, bear close resemblance to similarly-aged marsupial fossils from South America. Now the Australian continent is home to around 70 percent of all living marsupial species.

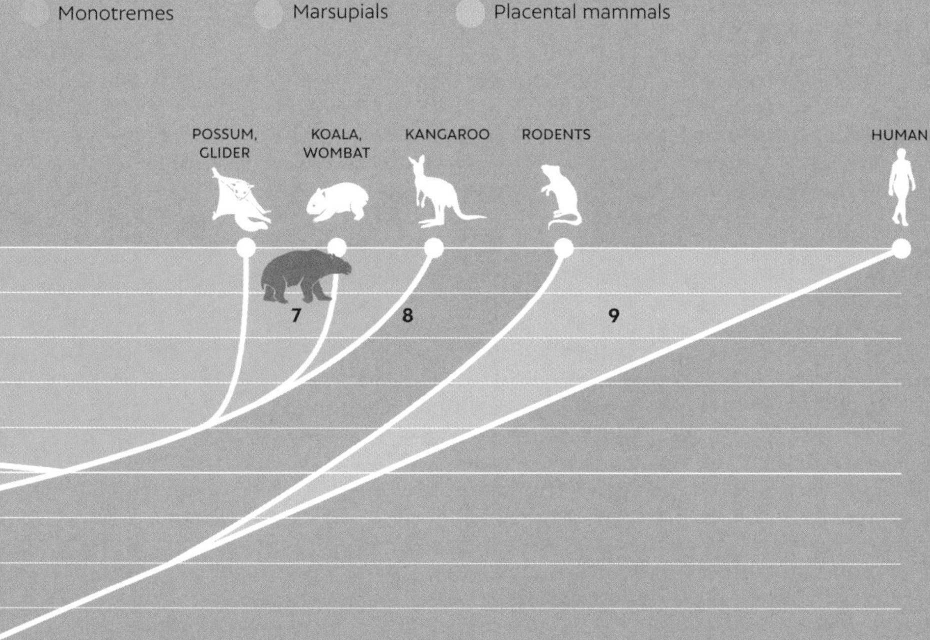

1 Mammals and reptiles go their separate ways.

2 Monotremes diverge from placental mammals.

3 Marsupials split from placental mammals.

4 *Teinolophos* was a tiny, platypus-like creature. It is known from just four pieces of jawbone, found in Australia.

5 One of the oldest-known marsupial fossils belongs to a mouse-size creature that climbed trees and ate insects, called *Sinodelphys*. It comes from China.

6 A shrew-like creature called *Juramaia sinensis*, also from China, is one of the earliest-known placental mammals.

7 *Diprotodon optatum* was the biggest marsupial known to have existed. The size of a hippo, it became extinct circa 25,000 years ago.

8 Australia has had kangaroos for at least 20 million years. *Balbarids*, which are a hopping, climbing distant cousin, became extinct circa 12 million years ago.

9 Placental mammals diversify.

DOMESTICATED BLISS

Domestication is one of our species' greatest inventions. The ability to domesticate animals and plants led our ancestors to abandon their hunter-gatherer lifestyles and settle down. It led to the emergence of villages and predictable food supplies, driving the development of trade, technology, agriculture, and urbanization. In short, domestication helped to fuel the rise of civilization.

Domestication is not the act of taming a wild animal. It refers instead to a process, hewn across multiple generations, in which humans gain increasing control of the biology and behavior of another species. It is underpinned by genetic changes and refined through the act of selective breeding; key individuals with choice characteristics are deliberately bred together.

CATTLE

All cattle are descendants of the aurochs: a belligerent, big-horned bovine that stood up to 6 feet (180 cm) tall at the shoulder. Its heyday was during the last Ice Age, when it was featured in cave paintings. Habitat loss and hunting pushed it to extinction, and the last aurochs died in 1627.

SHEEP (Near East)

GOAT (Middle East)

EUROPEAN COW (Near East)

A timeline of domestication and where it occurred

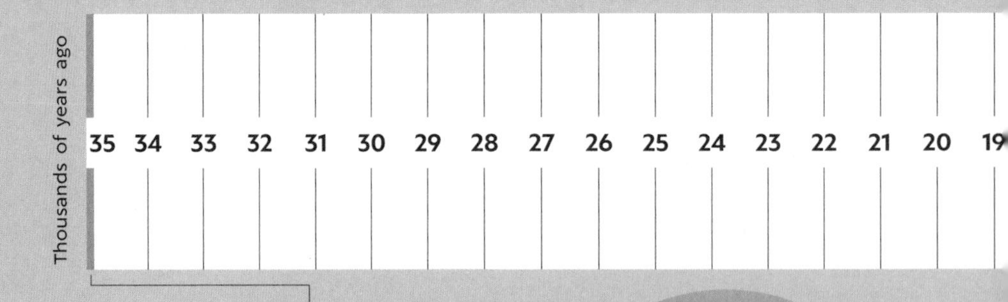

DOG (Eurasia)

DOGS

Dogs were the first animals to be domesticated. All dogs today, from tiny chihuahuas to enormous Great Danes, are descended from the same gray wolf ancestor. Now dogs and wolves share over 99.5 percent of their DNA, but the small difference is enough to imbue them with their vastly different characteristics.

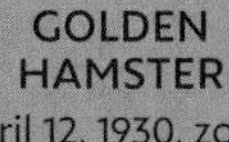

GOLDEN HAMSTER

On April 12, 1930, zoologist Israel Aharoni discovered a hamster nest in a Syrian field. He took the golden or Syrian hamsters to his lab in Jerusalem. Five escaped and some ate the others. But over the following 18 years, the remaining hamsters were selectively bred together and the population grew. By 1948, domestic hamsters were well established.

GEESE

Geese may have been the first birds to be domesticated, according to a 2022 study. Fossil remains from China hint that geese were either partially or fully domesticated 7,000 years ago.

LLAMA
(South America)

GOLDEN HAMSTER
(Western Asia)

GOOSE
(East Asia)

WESTERN HONEYBEE
(Middle East)

HUMPED COW
(South Asia)

PIG
(Near East)

HORSE
(Eurasia)

17 16 15 14 13 12 11 10 9 8 7 6 5 4 3 2 1 0

WHEAT AND BARLEY
(Middle East)

CATS

Domestic cats are descended from the African wildcat. It's thought that the famously independent creatures effectively domesticated themselves. When farming took off, around 10,000 years ago, populations of mice and other vermin exploded, and humans found cats to be useful as pest control.

SILK MOTH
(East Asia)

GOLDFISH
(East Asia)

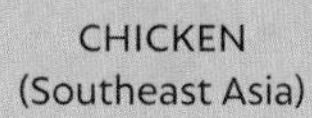

CHICKEN
(Southeast Asia)

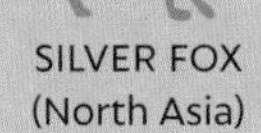

SILVER FOX
(North Asia)

CAT
(Middle East)

THE SILVER FOX

Domestication doesn't have to take thousands of years. In 1959, Russian scientist Dmitri Belyaev began an experiment that shows just how quickly it can happen.

He brought 130 wild silver foxes into captivity. Most were quite ferocious, but a few were slightly less aggressive. He bred these "friendly" animals together, and then repeated the process. Every generation, the friendliest foxes were paired together.

Over time, the foxes changed. They looked and acted differently, and became more doglike. What's interesting is that Belyaev didn't set out to create a doglike creature. Instead, canine-like characteristics emerged by selecting purely for friendliness. It took less than a human lifetime to almost completely domesticate the silver fox. The chart below is an estimation of the changes over generations and time.

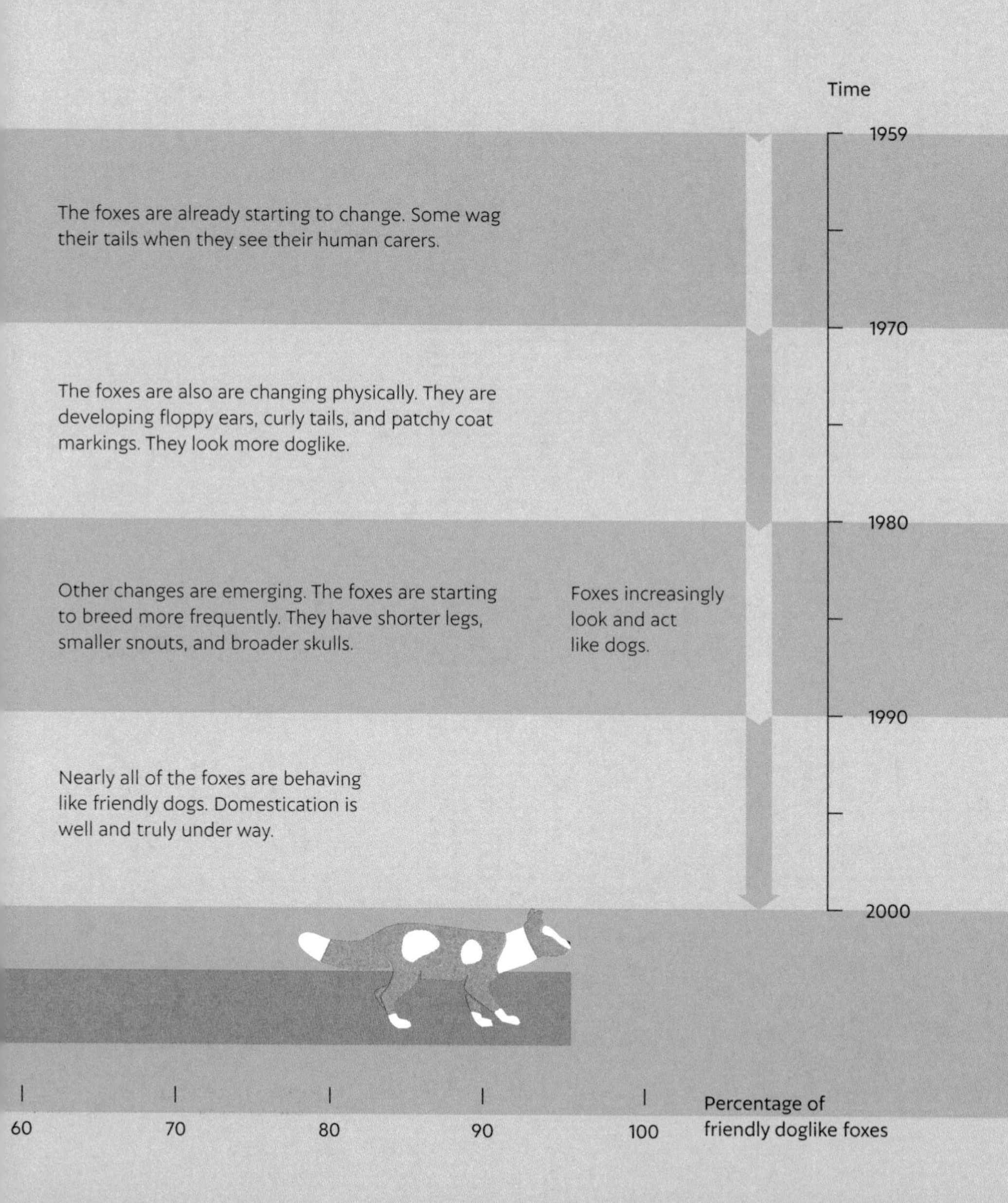

2
ECOLOGICAL SPANS

INTRODUCTION

Every organism is part of an ecosystem: a biological community of living, interacting organisms and the physical environment in which they live. Sometimes, these ecosystems are small, like a garden pond or a rock pool, but they can also be vast, such as the enormous tundra ecosystems north of the Arctic Circle or the lush, green rain forests scattered around the equator.

At first glance, these ecosystems look static. We see a snapshot of an ecosystem frozen in time, but the reality is that ecosystems are dynamic entities that are always in a state of flux. Organisms live and organisms die. Populations wax and wane. Species evolve and become extinct. Seasons change, and as the Earth rotates on its axis, day follows night follows day. Tides come in and go out again. Climate patterns change, as does the physical environment. Wind, water, and ice erode and shape the land.

It is for these reasons that ecological spans are incredibly varied. Life in intertidal rock pools, for example, experiences extreme change over the span of a single day. Ecology teaches us that all life on Earth is interconnected. We rely on our planet's rich and varied ecosystems for the air we breathe, the water we drink, the food we eat, and much more besides. Ecosystems aren't just nice to have. They're essential for life as we know it, and the decline of species and destruction of ecosystems caused by human behavior is a global problem.

The good news is that, provided they're not too badly pummeled, ecosystems can bounce back. Studies of life rising from the ashes of forest fires show us this, as does the pioneering work of conservationists in Brazil who are working to replant the damaged Atlantic Forest. With protective strategies in place, populations of endangered species can grow, and formerly imperiled species can once again blossom in the wild. These are some of the most poignant and rewarding time spans to witness.

Beavers are ecosystem engineers. They construct and maintain complex ecosystems that enable many other species to thrive.

WHALE DECOMPOSITION

BLOATING: A few hours or days
The body expands with gas as decomposition begins. It may float for a while. Sharks and seabirds dig in.

SINKING: Days to weeks
The corpse starts to sink. Mobile scavengers such as hagfish and sleeper sharks take chunks of blubber on the way down. The carcass comes to rest on the seafloor. It is now called a "whale fall."

WHALE FALL
There are four stages to a whale fall. These stages vary in length, and often overlap. They are shown below from head (Stage 1) to tail (Stage 4).

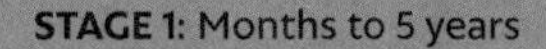

STAGE 1: Months to 5 years

MOBILE SCAVENGING
Deep-sea scavengers, including grenadier fish, lobster, and octopuses, join in the feast. They come from far and wide.

STAGE 2: Months to 2 years

ENRICHMENT-OPPORTUNIST
Any remaining soft tissue is eaten. Crustaceans and worms colonize the bones and the nutrient-dense sediments that surround the carcass.

WHAT HAPPENS TO A WHALE WHEN IT DIES?

Reaching lengths of 98 feet (30 m) or more, the blue whale is the largest animal ever known to have existed. Individuals can live for up to 90 years, but when they die, their remains can nourish neighboring life-forms for an equivalent time span. The whale's body provides an intense, sudden burst of nutrients to the creatures of the deep, and a whole new ecosystem develops around the corpse.

Scientists estimate that around 700,000 carcasses of the nine largest whale species in the world are decomposing at any one time. Each may nourish up to 400 different species. It's not easy to find or examine whales after they have died. Instead, scientists sometimes intentionally sink dead stranded whales so that they can study them.

STAGE 3: Up to 100 years

BONE-BUSTING

Bacteria, zombie worms, clams, and other organisms break down fat trapped in the bones. Sulphide chemicals are released, which other organisms can consume.

STAGE 4: Unknown length

REEF STAGE

The remaining bone fragments, which lack organic material but still contain minerals, are colonized by suspension feeders such as mollusks.

THE BENEFITS OF BURIAL AT SEA

Before industrial whaling began in the eleventh century, it is estimated that whale populations sunk between 190,000 and 1.9 million tons of carbon per year to the bottom of the ocean. That is the equivalent of taking between 40,000 and 410,000 cars off the road every year. When whales are killed and processed, however, instead of dying and sinking naturally, carbon is released into the atmosphere. It is estimated that twentieth-century whaling resulted in the emission of an additional 70 million tons of carbon dioxide. Whales can help in the fight against climate change, if only humans would let them live and die naturally.

The deep seafloor is cold, dark, and largely barren. The organisms that live there survive mainly on dead and decaying material that falls from the surface, including dead animals and plants, sand, soot, and fecal matter, which are rich in carbon and nitrogen. It's called "marine snow" because the falling flakes look a little like snowflakes. When flakes reach the bottom, they join the muddy sludge that covers it. Around three-quarters of the deep seafloor is covered in this sludge.

In this nutrient-scarce environment, a whale fall is like an all-you-can-eat buffet. A single decomposing whale provides the same amount of carbon as 2,000 years worth of marine snow falling onto a 538-square-foot (50 sq m) patch of seafloor.

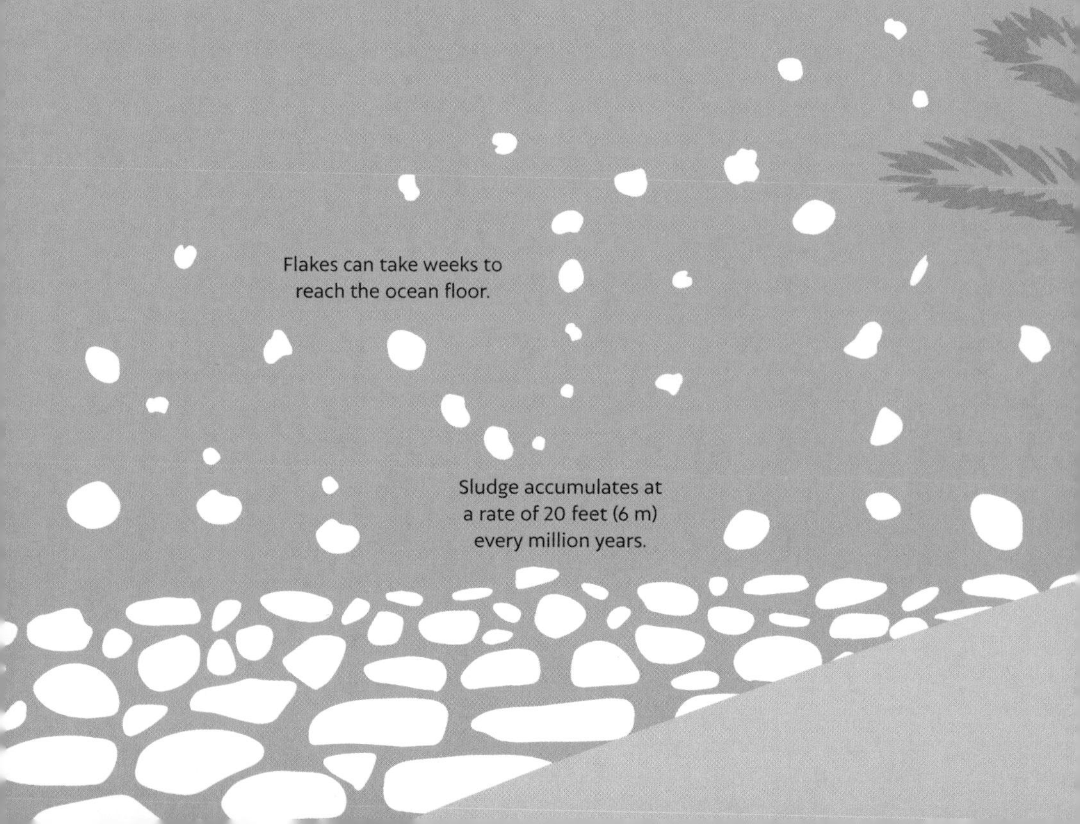

Flakes can take weeks to reach the ocean floor.

Sludge accumulates at a rate of 20 feet (6 m) every million years.

ZOMBIE WORMS

Zombie worms were first discovered in 2002, on a 2-mile-deep (3 km) whale fall in California's Monterey Bay. Unlike regular worms, they don't have a mouth or gut. Instead, they survive by producing specialized root tissues that bore into bone. At least 26 species are known, including *Osedax mucofloris*, which means "bone-eating snot-flower." Its "flowers" are feathery plumes of tissue, which waft in the water and absorb oxygen. Newborn larvae can survive for 10 days without eating, which gives them time to drift around and colonize other whale falls.

PIONEER SPECIES

0 YEARS

BARREN
In the immediate aftermath of a fire, ground looks barren but holds the seeds of life.

1–2 YEARS

ANNUAL PLANTS
The first arrivals are short-lived annuals, some of which disperse seeds to continue the species.

3–4 YEARS

GRASSES AND PERENNIALS
Hardier species take root and may be present all year or die off and regrow in the spring.

OUT OF THE ASHES

Even after a major disturbance, such as a devastating fire or flood, a landscape is seldom lifeless. The event may destroy trees and plants, drive out wildlife, and set the area back to an earlier biological stage, but its soil will retain nutrients and seeds. A process called "secondary succession" begins as the landscape regenerates and new flora and fauna emerge. Species, such as the North American jack pine, have a head start because they have evolved to adapt to fire. This tree needs a wildfire's heat to open its cones and release its seeds, while other pines have developed thicker, fire-resistant bark. In clearings, where trees have fallen, new ecosystems begin to develop, as plants that languished in the shade take advantage of the greater light and attract new species of insects to feed on them.

FIRE ECOLOGY

In nature, fire can be a potent force for good. Moderate, seasonal wildfires revitalize coniferous forest, shrubland, and grassland ecosystems. Fire renews by removing dead organic matter, returning nutrients to the soil, and stimulating new growth. It encourages biodiversity as plants and animals recolonize clearings. Some trees even need its heat to disperse their seeds. While controlled fires can help this process and limit rampant wildfires, suppressing all fire can drastically reduce biodiversity. Scientists now recognize that, in many landscapes, fire is a "natural disturbance" that is required to promote new life.

AUSTRALIA'S LOSSES

Fueled by extreme heat, drought, and strong winds, bushfires devastated vast areas of Australia in 2019. At least 1 billion creatures died, many of them species unique to the continent, such as the long-footed potoroo, mountain pygmy possum, and western ground parrot. Indigenous trees are better equipped to survive. Eucalyptus needs heat to open its seedpods, and other species store seeds in the soil or high in their crowns or they resprout from shoots under their bark.

INTERMEDIATE SPECIES

5–150 YEARS

GRASSES, SHRUBS, PINES, YOUNG OAK, AND HICKORY

These perennial species add undergrowth at first, but over years gain height, attracting birds and providing substantial shelter for a forest's wildlife.

CLIMAX COMMUNITY

150+ YEARS

MATURE OAK AND HICKORY FOREST

The tallest, oldest trees are called the "climax community," as this is the final stage of a forest's development. They endure unless chopped down or destroyed by fire.

BEAVERS: ECOSYSTEM ENGINEERS

Beavers are remarkable animals with their chisel-like incisors, webbed back feet, and flat, scaly tails, and they are essential in helping other species to thrive. Beavers profoundly change their landscape and boost local biodiversity by building dams and lodges. It's been suggested that beavers can be used as a way to improve the health of degraded farmland and other impoverished habitats.

Around 20 years ago, scientists working with Scottish National Heritage in the UK began an experiment to see what would happen if beavers were released into a 2-square-mile (5 sq km) plot of disused farmland. The site had been drained in the 1800s. By 2000, it contained grassland and small clusters of deciduous trees, such as willow and alder. It also had a drainage ditch, which was supplied by a local spring.

How beavers change a landscape

2002

A pair of beavers were released and began to fell trees and build their first dam. It was 10 feet (3 m) long and raised the stream level by 2.3 feet (70 cm).

2006

The beavers started to breed. The average number of beavers present during the study was four.

The landscape was transformed by the beavers. They created more than 640 feet (195 m) of dams, 1,640 feet (500 m) of canals, and an acre of ponds. In turn, this created a variety of new environments, including freshwater pools and marshland.

During this time, the total number of species increased by approximately 150 percent, boosting a varied mix of plants and animals. From dams to damselflies, the setting provided new habitat for plants, invertebrates, amphibians, fish, birds, and other mammals.

Further studies have shown how beavers and their dams can also help to improve water quality, prevent soil erosion, and minimize flooding.

2016

The beavers kick-started the regenerative process, providing homes and opportunities for all of the diverse wildlife that returned.

A MASTER BUILDER

Beavers are the undisputed lumberjacks of the animal kingdom. Their ability to create new habitats with the trees they cut down is so impressive, that beavers are known as "ecosystem engineers."

How beavers build their dams

5 MINUTES

With its sharp teeth and powerful jaws, an adult beaver can fell a relatively young sapling in as little as five minutes.

1 NIGHT

In one night, an adult beaver can cut down a 65-foot (20 m) poplar tree, divide it into a dozen sections, and then drag the logs to the water. An adult beaver can carry its own bodyweight in logs.

< 1 DAY

In less than one day, beavers can build a watertight dam. They start by driving branches and logs into the mud at the bottom of a stream. Sticks, bark, rocks, mud, and pieces of plants are then used to create the main structure. The upstream water level begins to rise and a pond is formed. When the water level reaches around 2.5 feet (80 cm), it becomes deep enough to house the beaver's lodge.

BEAVERING AWAY

There are two species of beaver: the North American beaver and the Eurasian beaver. Overhunting in the nineteenth and early twentieth century almost drove them to extinction, but now both species are doing well, and beavers continue to be reintroduced to many parts of Europe.

~ 2 NIGHTS

In a couple of nights, beavers can build a rudimentary lodge that is strong enough to withstand the oncoming winter.

~ 2 WEEKS

In two weeks, the same lodge can be elaborated into a truly grand design. Beaver lodges can be 39 feet (12 m) across and more than 10 feet (3 m) tall. They are made from carefully placed sticks, logs, and vegetation. Mud is used to bind the walls together and insulate against the cold, but a little air vent is left at the top. The lodge contains three separate rooms: a nursery, a bedroom, and a feeding chamber, and it also has several concealed underwater entrances.

Across decades, a single beaver lodge can house generations of beavers, and at any one time, there may be multiple generations of the same family living in the same abode.

THE LUNGS OF THE WORLD

Tropical rain forests are often called the "lungs of the world" because they breathe in carbon dioxide and breathe out oxygen, but they do so much more. They help stabilize the world's climate and maintain its water cycle, and they are home to a staggering amount of biodiversity.

Deforestation in South America: then and now

AMAZON RAIN FOREST
The Amazon rain forest is the world's largest and best-known rain forest. It has been teeming with life for more than 55 million years, and is currently home to one in 10 of all known species. About 17 percent has been destroyed in the last 50 years, and scientists now fear the Amazon rain forest is reaching a tipping point where the remaining trees could die off en masse.

HARPY EAGLE
The enormous harpy eagle is becoming increasingly rare in the Amazon rain forest, and is all but extinct in the Atlantic Forest. The decline is driven primarily by habitat loss, as the apex predator struggles to hunt in open areas, and the remaining pockets of forest are too small to sustain it. Its conservation status is currently listed as vulnerable.

Some 30 million species of plants and animals call the rain forest their home, and yet, this ecosystem is under threat. Tropical rain forests are being cut down to make way for cattle ranches, plantations, and mines. More than 3,860 square miles (10,000 sq km) are destroyed every year.

Original forest area

Deforested area

Current forest cover

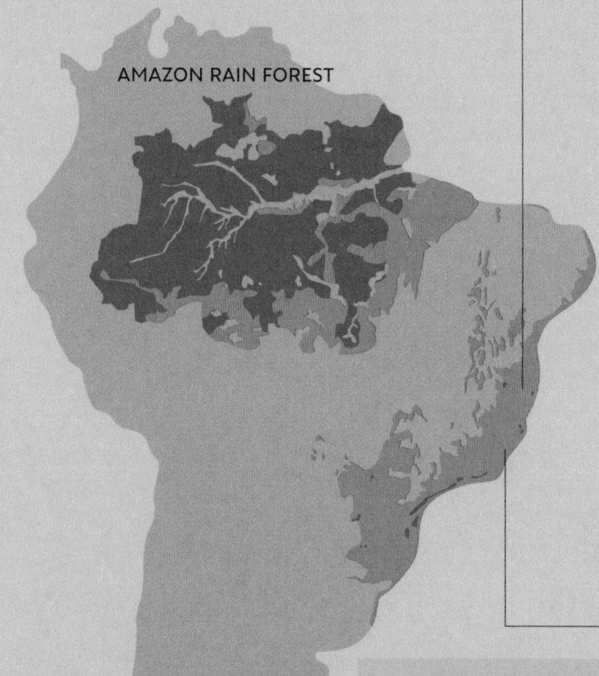

ATLANTIC FOREST

The Atlantic Forest may be less well known than the Amazon rain forest, but it is just as important. Around one in 50 of every plant and vertebrate land animal species lives here and nowhere else. It is made up of a patchwork of different ecosystems, including tropical rain forest, cloud forest, and grassland. More than 80 percent has been destroyed over the last 500 years, and much of what remains is in small, unconnected fragments. Now scientists think that some Atlantic Forest ecosystems could collapse within 50 years.

BUFFY-TUFTED MARMOSET

With its distinctive facial markings, the buffy-tufted marmoset is also called *sagui caveirinha*, or the "little skull monkey." It's found only in the mountainous regions of the Atlantic Forest, where habitat loss and fragmentation are driving its decline. Numbers have halved in just 18 years (three generations). Its conservation status is now listed as endangered.

HOW TO REGROW A RAIN FOREST

Although rain forests continue to be cut down, there are glimpses of good news. When deforested areas are left alone, they can regrow. Regenerated forests like this are called "secondary forests."

About one-third of Brazil's lost rain forests are recovering naturally. This is offsetting around 12 percent of the total carbon emissions from the Amazon's deforestation. However, this cannot be an excuse to cut down primary forest.

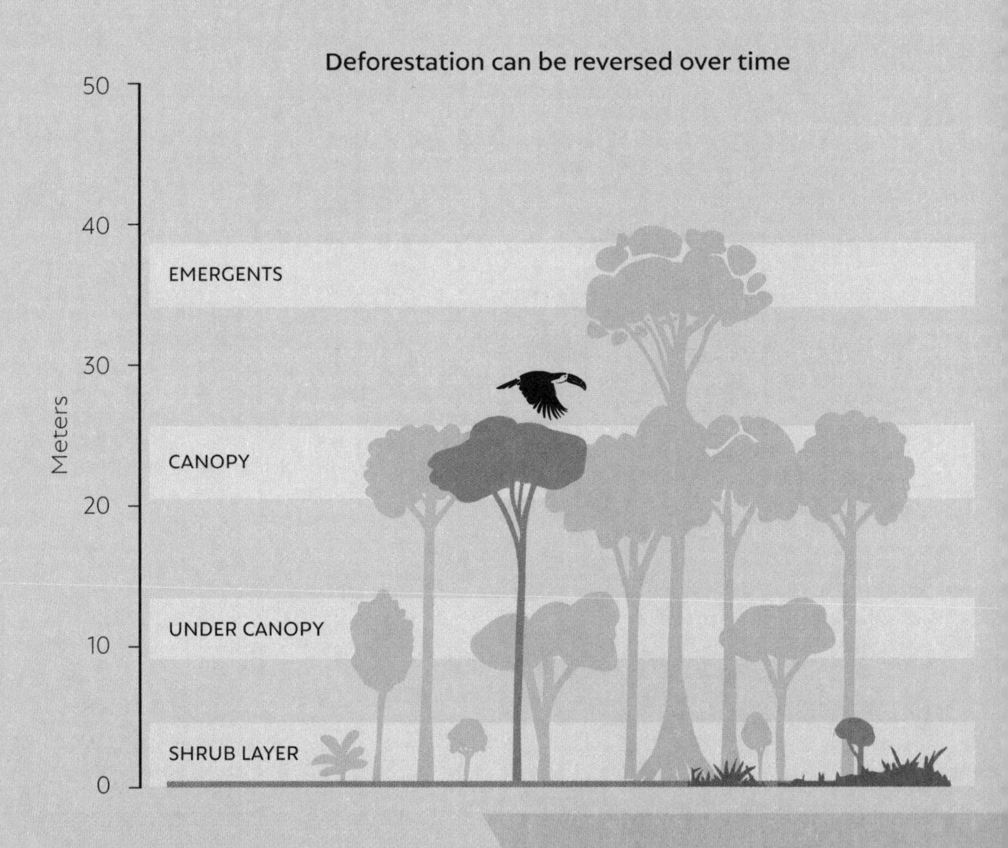

0

Growth starts immediately but takes a year or two to really get going.

15 YEARS

It takes 15 years for this new growth to resemble a forest.

BRAZIL'S EARTH INSTITUTE

With a helping hand, secondary forests can be encouraged to grow more quickly. At Brazil's Earth Institute—a former cattle ranch—locals planted more than 2.5 million seedlings of 297 Atlantic Forest species over 15 years. Now the rain forest is back, and it is steadily being recolonized by iconic species such as the maned wolf, the jaguar, and the red-crowned parrot.

It takes 40 years for these secondary forests to recover 85 percent of their original biodiversity. However, this can only occur if there are animals and plants left that can recolonize and continue to breed and grow.

FLOUNDERING FROGS

Frogs and toads play a key role in many ecosystems. They are both predators and prey, helping to keep delicate food chains in balance. Today, however, they find themselves in the midst of an amphibian apocalypse. A deadly fungus is spreading around the globe. It causes a disease called "chytridiomycosis" and is responsible for the greatest recorded loss of biodiversity due to disease.

Chytridiomycosis is caused by two species of fungus. The fungus infects and damages the skin, impairing the amphibian's ability to breathe and regulate its water levels. This can lead to a heart attack. If the conditions are right, the fungus can survive outside its host.

Timeline of a frog-killing fungus

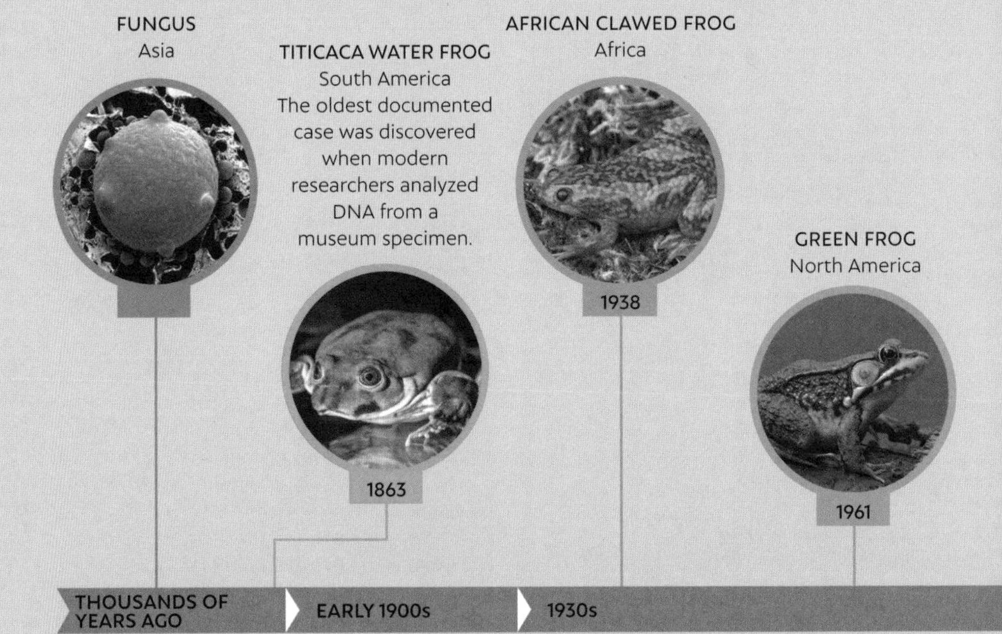

The fungus existed thousands of years ago, but it only caused disease on a local basis. It probably originated in Asia.

Modern analyses of museum frogs show that the chytrid fungus was present in southern Africa and Asia in the early 1900s, where it probably caused small-scale outbreaks.

The start of the international trade in amphibians. Frogs were shipped around the world for use in medical research and as pets. The disease spreads with them.

THE INTERCONTINENTAL SPREAD OF THE CHYTRID FUNGUS

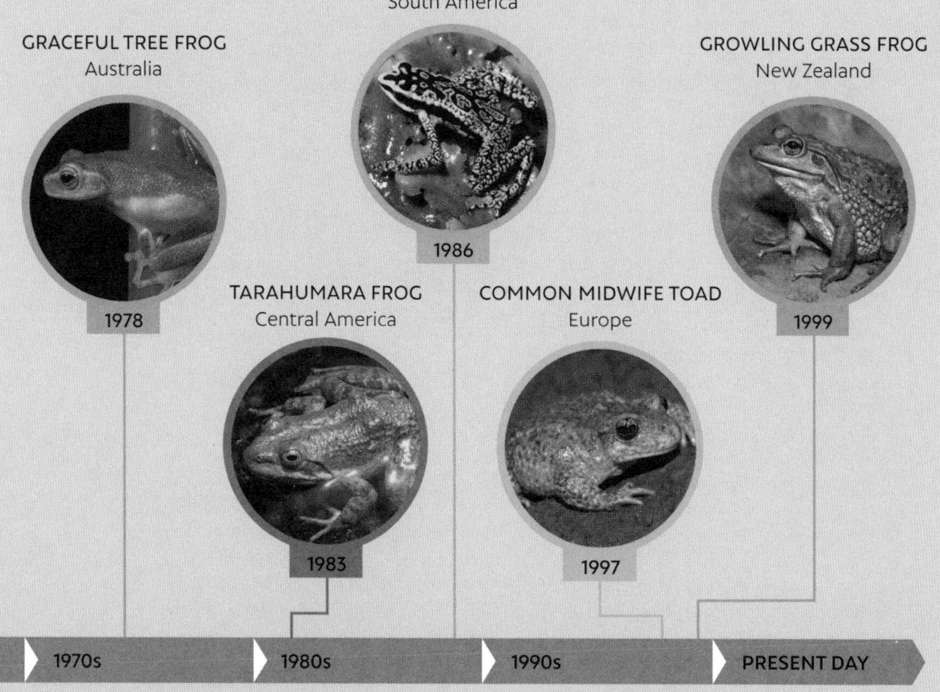

1970s

Scientists noticed that frogs were dying. The chytrid fungus caused the decline of hundreds of populations of frogs in Australia and the Americas.

1980s

The fungus went global. Frogs declined and deaths were at their worst, but frogs in Asia were spared. They had resistance to the disease.

1990s

Scientists in Australia identified the chytrid fungus, but it had already caused the extinction of at least four frog species.

PRESENT DAY

The chytrid fungus has caused the decline of at least 500 amphibian species, including 90 presumed extinctions. Only 12 percent of affected species are showing signs of recovery. The fungus remains a major threat globally.

LIFE BETWEEN THE TIDES

Tides ebb and flow twice per tidal day. A tidal day lasts 24 hours and 50 minutes and is different from a solar day, which is shorter at only 24 hours. A tidal day is also called a "lunar day," because it is the time taken by the moon to complete one rotation on its own axis. As tides ebb and flow, the landscape is alternately transformed, from one of swirling seawater to one of parched rocks. Intertidal zones are found anywhere the ocean meets the land, from rocky cliffs and coral reefs to mangroves and mudflats. The organisms that live here are adapted to these harsh extremes.

Intertidal wildlife changes with time

SPRAY ZONE

This area is splashed by ocean spray and high waves. It's only submerged during very high tides or severe storms.

MID TO HIGH ZONE

Life is less abundant here. Most residents have shells. They are either mobile, like hermit crabs that head for deeper water when the tide goes out, or immobile, like barnacles, which hide in their shells to avoid drying out and being eaten.

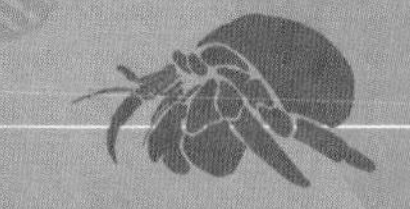

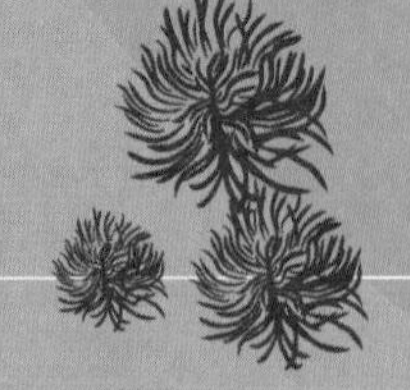

MID TO LOW ZONE

Life is more abundant here. Living things are less likely to dry out, so soft-bodied organisms do well. Many, such as anemones and seaweeds, are anchored to the rocks below, where they can withstand being battered by the waves.

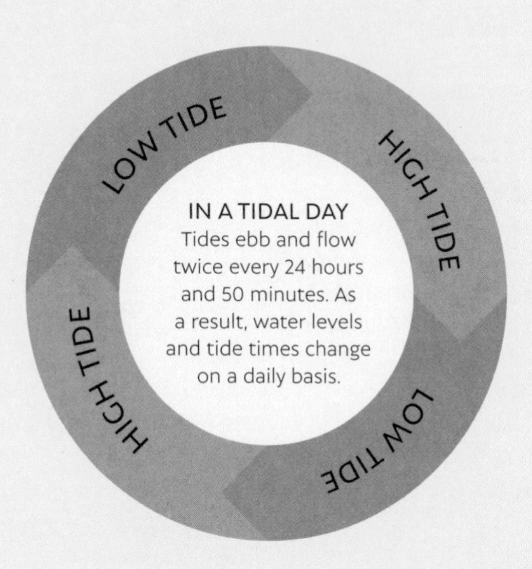

HIGH INTERTIDAL ZONE
This area is only covered by the highest of high tides and remains dry for long stretches of time.

MIDDLE INTERTIDAL ZONE
The area is regularly exposed and submerged, twice per tidal day, as tides ebb and flow.

SEAWEED

LOW INTERTIDAL ZONE
This area is almost always underwater, except during the lowest tides.

SEA SLUGS

STARFISH

MARINE ALGAE

GOING, GOING, GONE

When a species becomes extinct, it threatens the stability of the ecosystem in which it once lived. Imagine a giant tower made from wooden bricks. When a couple of bricks are removed, the tower might start to wobble. However, when too many bricks are removed, the tower will eventually collapse. There's a normal background rate of extinction (see below), but over the last few hundred years, the extinction rate has been increasing, and it's now thought that around 1 million animal and plant species could become extinct in the next few decades. Scientists are concerned this puts some ecosystems, and the services they provide, in danger.

THYLACINE

In just 200 years, the thylacine population declines from 5,000 to one. The thylacine, or Tasmanian tiger, was an apex predator and carnivorous marsupial that once lived in Australia, Tasmania, and New Guinea. The last-known thylacine, Benjamin, died alone in Tasmania's Beaumaris Zoo on September 7, 1936.

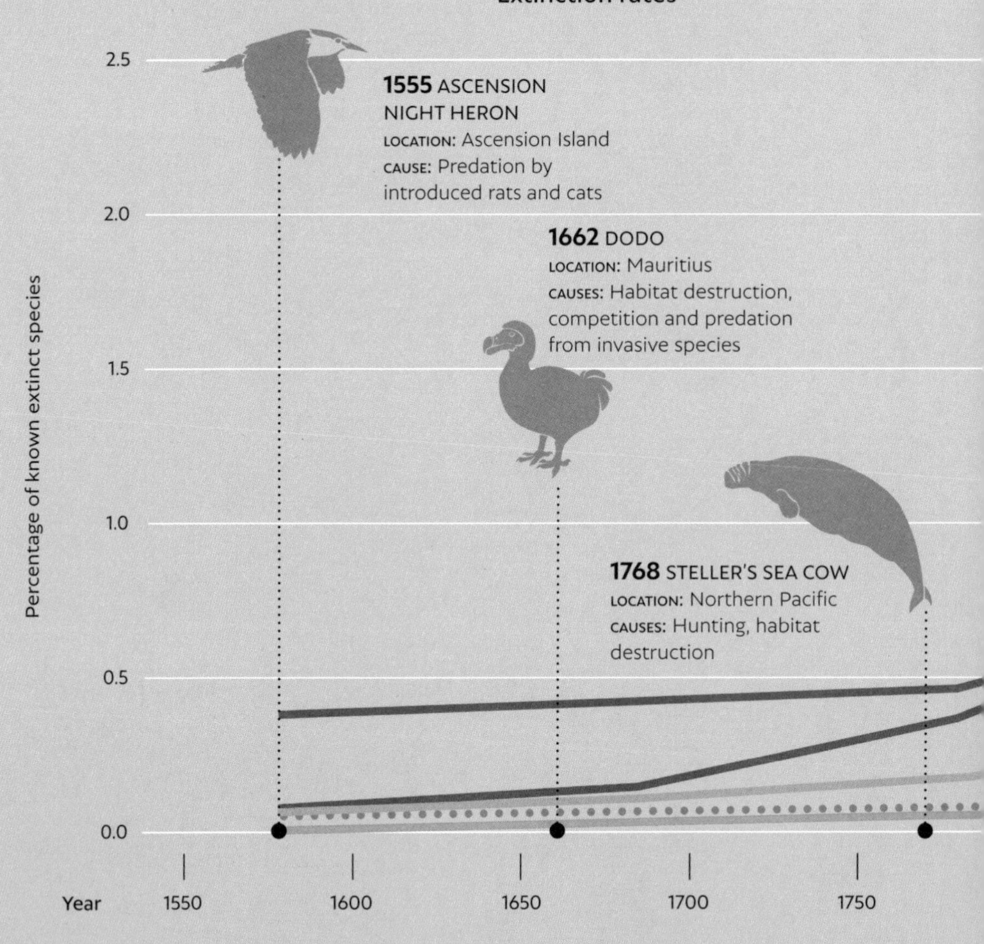

Today, more than 40 percent of amphibian species, almost 33 percent of reef-forming corals, and more than one-third of all marine mammals are threatened with extinction. Less is known about insects, but it is estimated that 10 percent are at risk. At least 680 vertebrate species have been driven to extinction since the sixteenth century.

PASSENGER PIGEON

In just 200 years, passenger pigeon populations decline from billions to one. They were once the most numerous bird in North America. They flocked in the millions. The last passenger pigeon, Martha, died alone in America's Cincinnati Zoo on September 1, 2014.

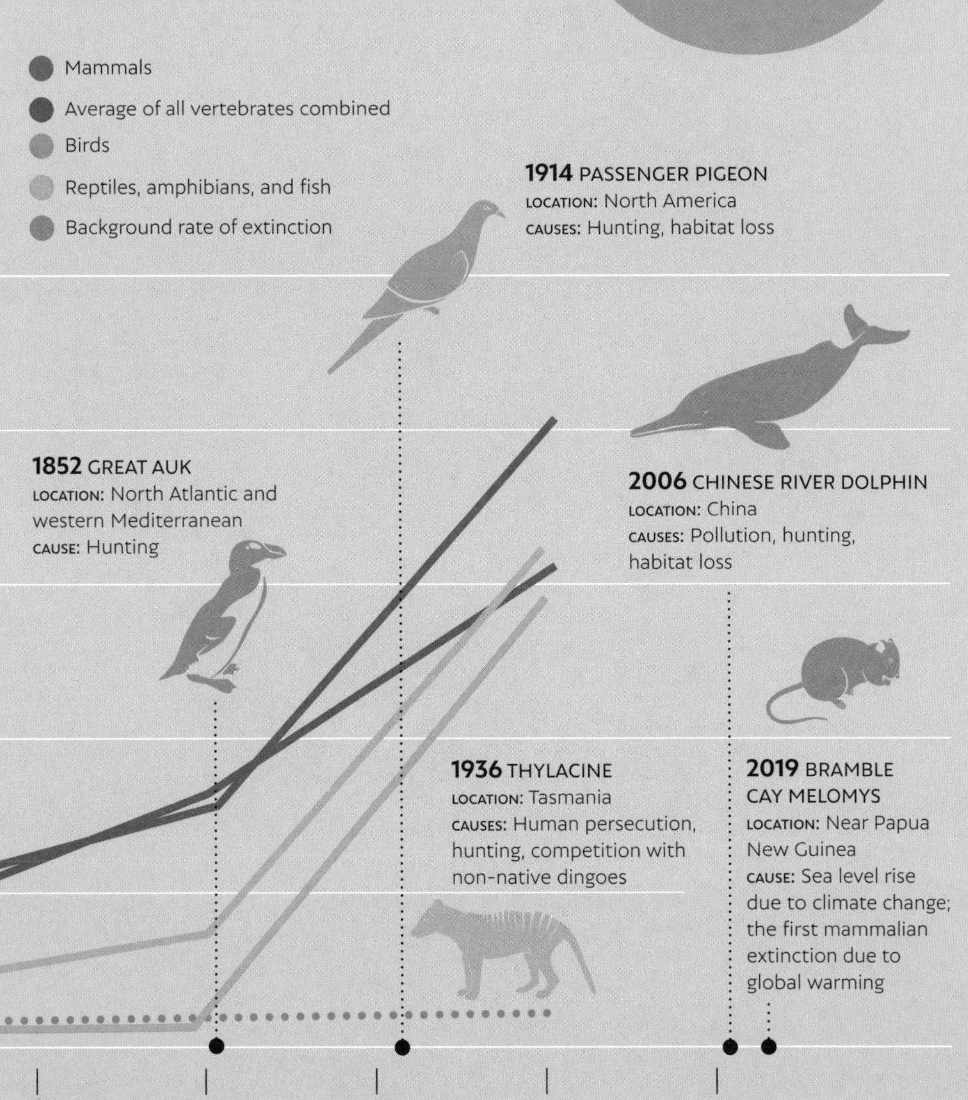

THE DEMISE OF STELLER'S SEA COW

All living things exist in complex food webs, so the disappearance of one species can influence the well-being and welfare of many other species. This ripple effect is known as a "trophic cascade." It means that extinction can breed extinction, and can occur over a surprisingly short time span, as the story of Steller's sea cow illustrates only too well.

Timeline of extinction

1741 Steller's sea cow is discovered by Georg Wilhelm Steller when his expedition becomes stranded on the Commander Islands. They kill the animal for food.

Over the next 27 years, other Alaska-bound sailors stop by and feast on sea cows. The sea cows' numbers start to decline.

Sea otters also live in the same area. They are hunted for their fur. Their numbers start to drop.

Sea otters eat sea urchins, and they eat the holdfasts that keep the kelp anchored to the ocean floor.

Steller's sea cow was an herbivorous marine mammal that lived in kelp forests, where it ate the kelp. It grew up to 30 feet (9 m) long and was related to the dugong and the manatee. However, its fate was ultimately tied to that of the sea otters, sea urchins, and kelp with which it coexisted.

1748 Sea otters decline due to hunting, which results in more kelp-munching sea urchins, and the kelp forest starts to collapse.

1753 Sea otters become extinct in this marine area. The kelp forest is further weakened.

Steller's sea cows feed primarily on kelp. Their numbers dwindle as they slowly start to starve.

1768 The demise of local sea otters triggers a fatal trophic cascade. Steller's sea cow becomes extinct.

CONSERVATION SUCCESS STORIES

Populations of many wild species may be dwindling, extinction rates may be spiking, and yet, there is still good news to be found. The modern conservation movement was founded in the late nineteenth century, and over 50 years ago, the US Endangered Species Act was passed. Under the Act, at-risk species can be highlighted, or "listed," and then protected and nurtured back to health. Species are then "delisted" when their situation improves.

The listing and delisting of key species

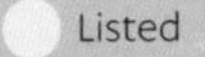

ROBBINS' CINQUEFOIL

BLACK-FOOTED FERRET

Populations plummeted when this North American mustelid became ill with the sylvatic plague. The species was declared extinct in 1979, but then a small population was found and used to establish a captive breeding program. Now more than 4,000 individuals have been released back into the wild, and captive breeding is ongoing.

HUMPBACK WHALE

This gentle giant was targeted by the whaling industry. In the 1960s, its population fell to around 5,000, but it began to recover after the ban on commercial whaling in 1986. There are now more than 100,000 individuals worldwide, but they continue to be affected by vessel collisions, habitat destruction, and other human-related problems.

MEXICAN CROCODILE

CALIFORNIA CONDOR

1970 1975 1980 1985

ROBBINS' CINQUEFOIL

Robbins' cinquefoil is a tiny, yellow-flowered perennial, which is found exclusively above the treeline of New Hampshire's White Mountains. It became imperiled when walkers started trampling on it. Less than 4,000 plants existed at the time of its listing, but this number has more than tripled, as awareness of the plants' plight continues to grow, and as a result footpaths have been rerouted.

LESSER LONG-NOSED BAT

MAGAZINE MOUNTAIN MIDDLE-TOOTHED SNAIL

LAKE ERIE WATER SNAKE

STELLER SEA LION

Human overfishing of this marine mammal's main food source—pollock and herring—contributed to its decline in Alaskan waters; populations have fallen by 75 percent since the 1970s. The situation improved after the fish stocks were managed and trawlers avoided the Steller sea lions' breeding sites.

SANTA CRUZ ISLAND FOX

1995 2000 2005 2010 2015 2020

SOARING TO RECOVERY

The California condor is a bald, beaked vulture and the largest North American land bird. Not so long ago, numbers plummeted and the species became extinct in the wild, and yet, conservationists have managed to pull it back from the brink.

OBSTACLES TO RECOVERY

Today, the wild condors still suffer from lead poisoning, so they are intermittently captured and treated with a drug to purge the lead from their bloodstream. Around 10 percent of the wild population are in veterinary respite at any one time.

The great California condor comeback

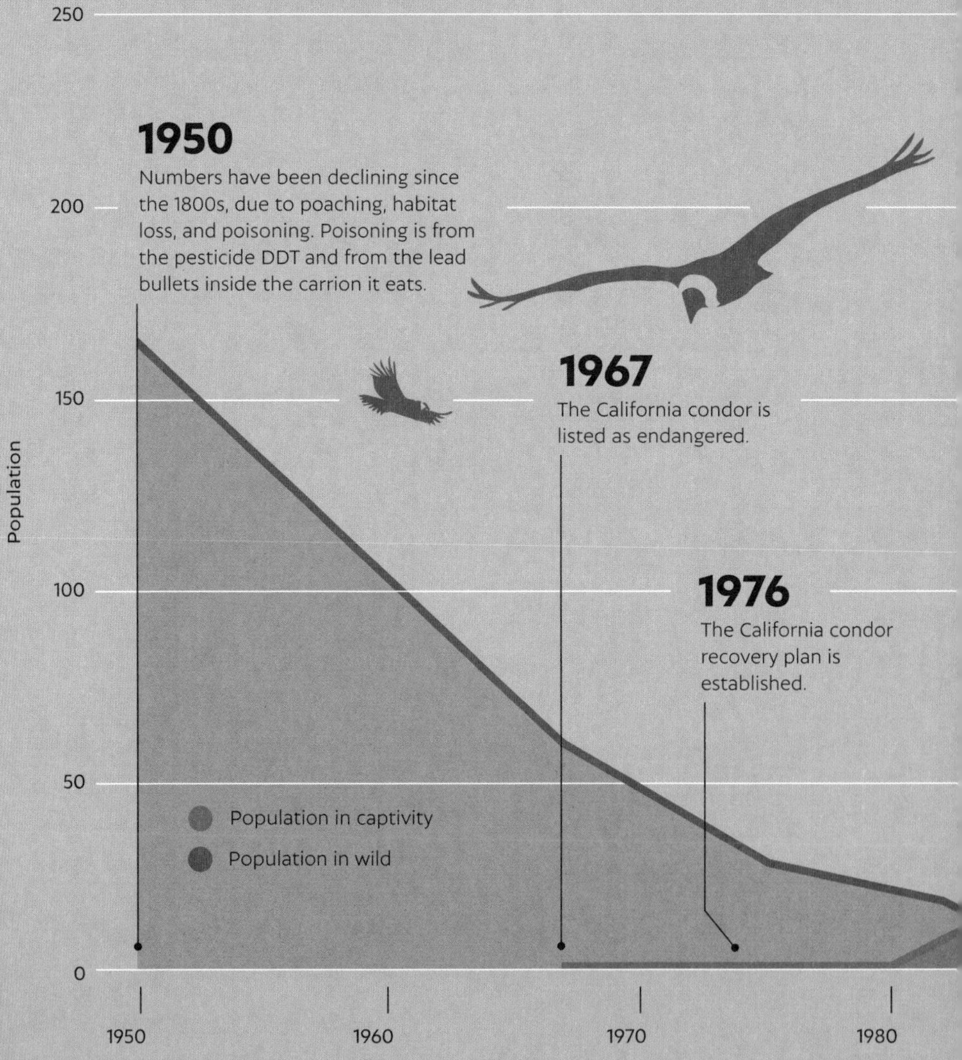

250

200

150

100

50

0

Population

1950 1960 1970 1980

1950

Numbers have been declining since the 1800s, due to poaching, habitat loss, and poisoning. Poisoning is from the pesticide DDT and from the lead bullets inside the carrion it eats.

1967

The California condor is listed as endangered.

1976

The California condor recovery plan is established.

Population in captivity

Population in wild

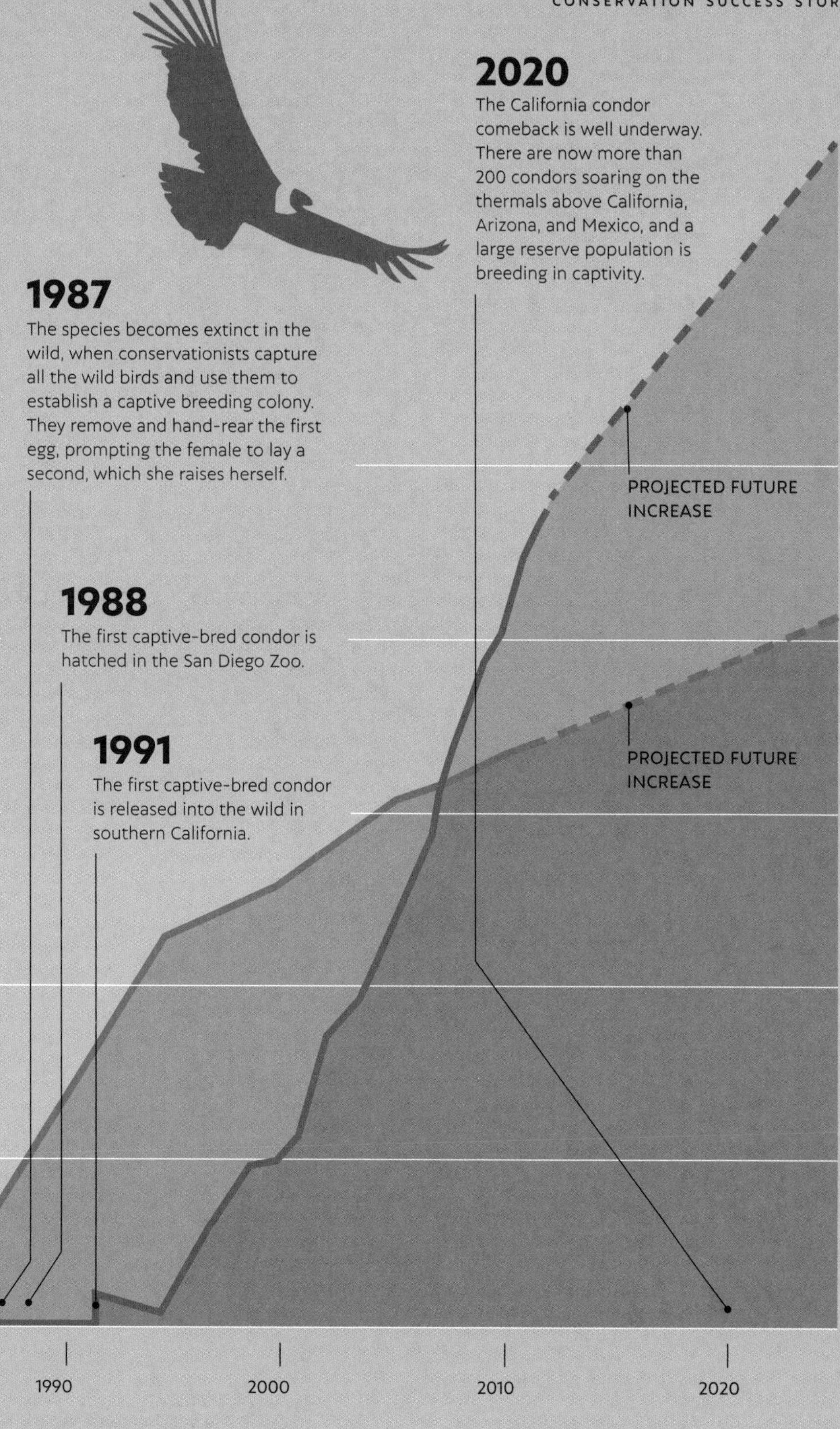
2020
The California condor comeback is well underway. There are now more than 200 condors soaring on the thermals above California, Arizona, and Mexico, and a large reserve population is breeding in captivity.
1987
The species becomes extinct in the wild, when conservationists capture all the wild birds and use them to establish a captive breeding colony. They remove and hand-rear the first egg, prompting the female to lay a second, which she raises herself.
PROJECTED FUTURE INCREASE
1988
The first captive-bred condor is hatched in the San Diego Zoo.
1991
The first captive-bred condor is released into the wild in southern California.
PROJECTED FUTURE INCREASE
1990
2000
2010
2020

3
LIFE SPANS

INTRODUCTION

We can all learn a thing or two from Creme Puff, the domestic cat owned by Jake Perry of Austin, Texas. She lived a full and active life, and enjoyed a varied diet that included cat food, broccoli, eggs, bacon, and coffee with cream. Every few days, she also enjoyed an eyedropper full of red wine, which, her owner claimed, circulated in her arteries and contributed to her long life. Creme Puff died when she was 38 years old.

This is an impressive age for a cat, yet there are many organisms that regularly exceed our expectations of longevity and survival. No one is quite sure how they achieve this, but genetics, environment, and lifestyle all play their part. The oldest individual tree is a 5,000-year-old bristlecone pine, while a colony of male quaking aspen trees has notched up tens of thousands of years. The oldest-known human is a woman from France, who died aged 122, while bowhead whales regularly live for a century or more.

At the other end of the extreme, adult mayflies have but a day on the wing before they meet their maker, while Müller's giant Sunda rat, a native of southeast Asia, is one of the shortest-living mammals on Earth, living for only 6–12 months. Death, it seems, catches up with us all, and yet, some organisms have evolved ways to delay the Grim Reaper.

Dormancy, practiced by some plants, animals, and microbes, provides a way to sit out hard times and then spring back to life. If you don't mind the wait, it can add millennia to the clock. Seeds, for example, have been revived after tens of thousands of years, while the Arctic permafrost continues to yield examples of multicellular organisms that have been frozen to near-death, then have thawed and survived.

The oldest-known tree in the world is a Great Basin bristlecone pine called Methuselah. It lives in the White Mountains of California.

HOW OLD?

The average life expectancy of a human being is 73 years old, although many of us will live for much longer. But it's nothing compared to certain organisms that live for hundreds or even thousands of years.

Life span is determined by multiple factors. Genetics plays a role, as does the environment. Some of the longest-living animals, such as glass sponges and tube worms, live in cold, deep waters, where they are stationary and have low metabolic rates. This slowing of metabolism may also slow the aging process, leading to longer life.

Size is important, too. Elephants, for example, tend to live longer than mice, which often outlive flies. Compared to bigger animals, smaller animals are more prone to predators, so over time, they often evolve to grow, reproduce, and die more quickly. Larger animals are better at fending off predators, so they take time to reach their adult size and reproduce many times through their life.

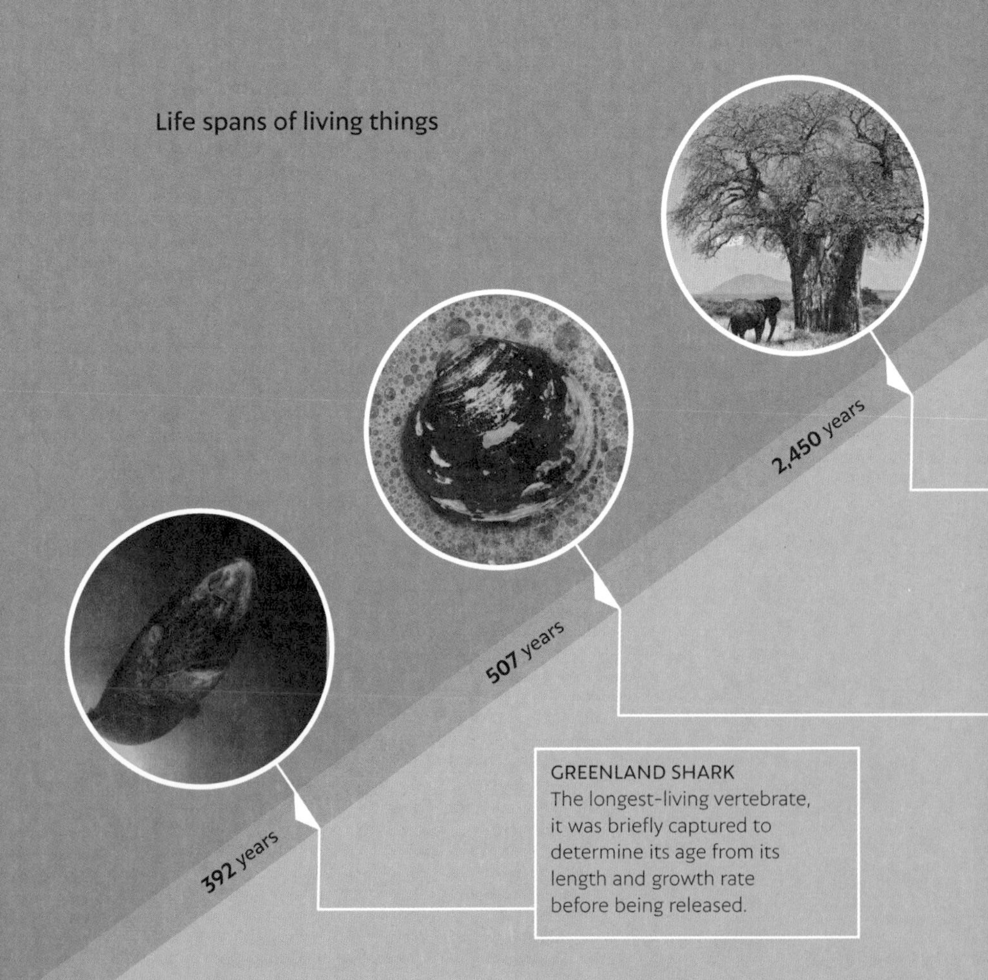

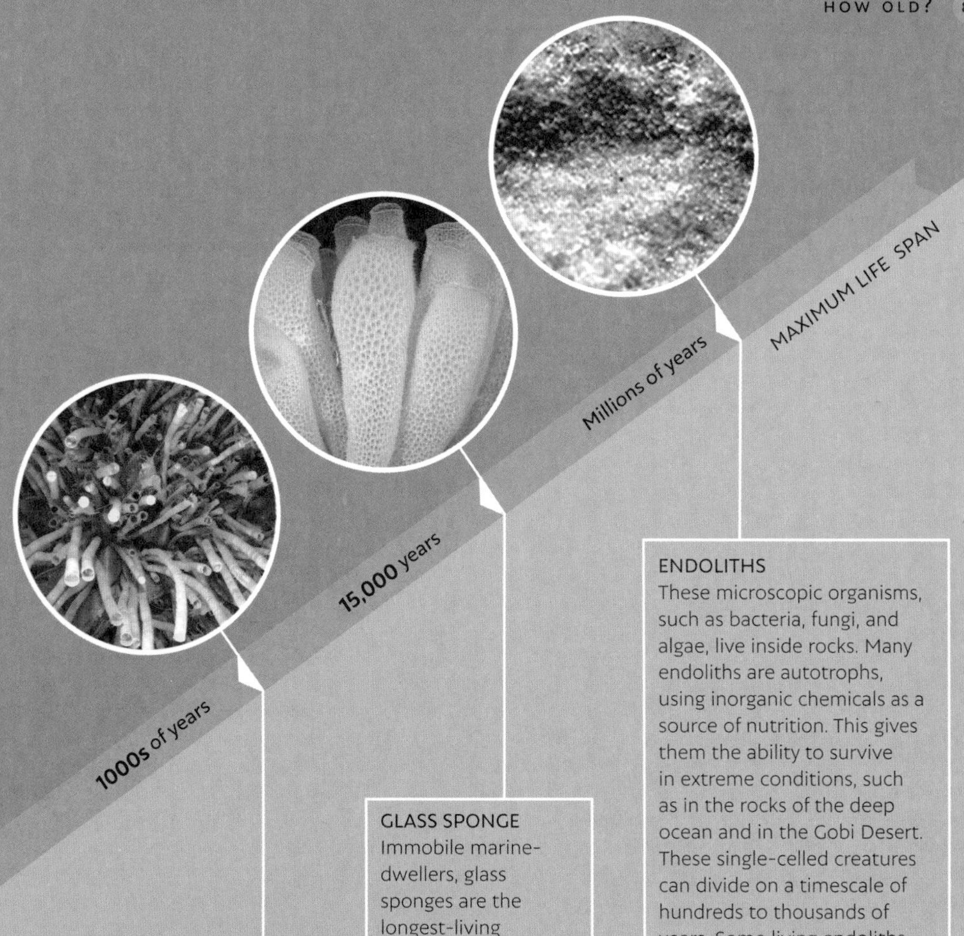

ENDOLITHS
These microscopic organisms, such as bacteria, fungi, and algae, live inside rocks. Many endoliths are autotrophs, using inorganic chemicals as a source of nutrition. This gives them the ability to survive in extreme conditions, such as in the rocks of the deep ocean and in the Gobi Desert. These single-celled creatures can divide on a timescale of hundreds to thousands of years. Some living endoliths may be millions of years old.

GLASS SPONGE
Immobile marine-dwellers, glass sponges are the longest-living multicellular animals on Earth.

PANKE BAOBAB
The oldest-known flowering plant, a Zimbabwean Panke baobab, died in 2011.

TUBE WORM
The tube worm *Escarpia laminata* lives in deep sea cold seeps. Individuals regularly live a few hundred years, but some are thought to live for millennia.

OCEAN QUAHOG
The oldest-known ocean quahog clam, called Ming, was dredged off Iceland's coast in 2006. Its age was estimated by counting the annual growth lines on its shell.

MAXIMUM LIFE SPAN

211 years

200 years

140 years

122 years

BOWHEAD WHALE
The oldest-known mammal

RED SEA URCHIN
This small, spiny invertebrate, which lives in shallow coastal waters, can live up to 200 years. It shows few signs of aging and continues to reproduce throughout its life.

LOBSTER
George, the oldest-known lobster, was captured off the coast of Newfoundland in 2008. He lived briefly in New York's City Crab and Seafood restaurant, until an animal rights group successfully lobbied for his freedom. He was released into an area where lobster fishing is banned. His age was calculated from his weight of 20 pounds (9 kg).

HUMAN
Jeanne Calment of France was one of the oldest-known humans. She was born in 1875 and died in 1997, age 122.

CAVE SALAMANDER
A cave salamander called "the olm" is the longest-living amphibian, but it is not the most obvious candidate for a long life. It's small, fragile, and eyeless, and lives a dark existence in the caves of the Dinaric Alps in southern Europe. Scientists attribute its long life to its unusually sluggish metabolism and largely predator-free environment.

ASIAN ELEPHANT
Dakshayani was the longest-living Asian elephant. She lived to the age of 89. In the wild, Asian elephants tend to live for around 50 years, but their life span can be greatly extended in captivity, such as for Dakshayani, who lived in an Indian temple.

Life spans of living things

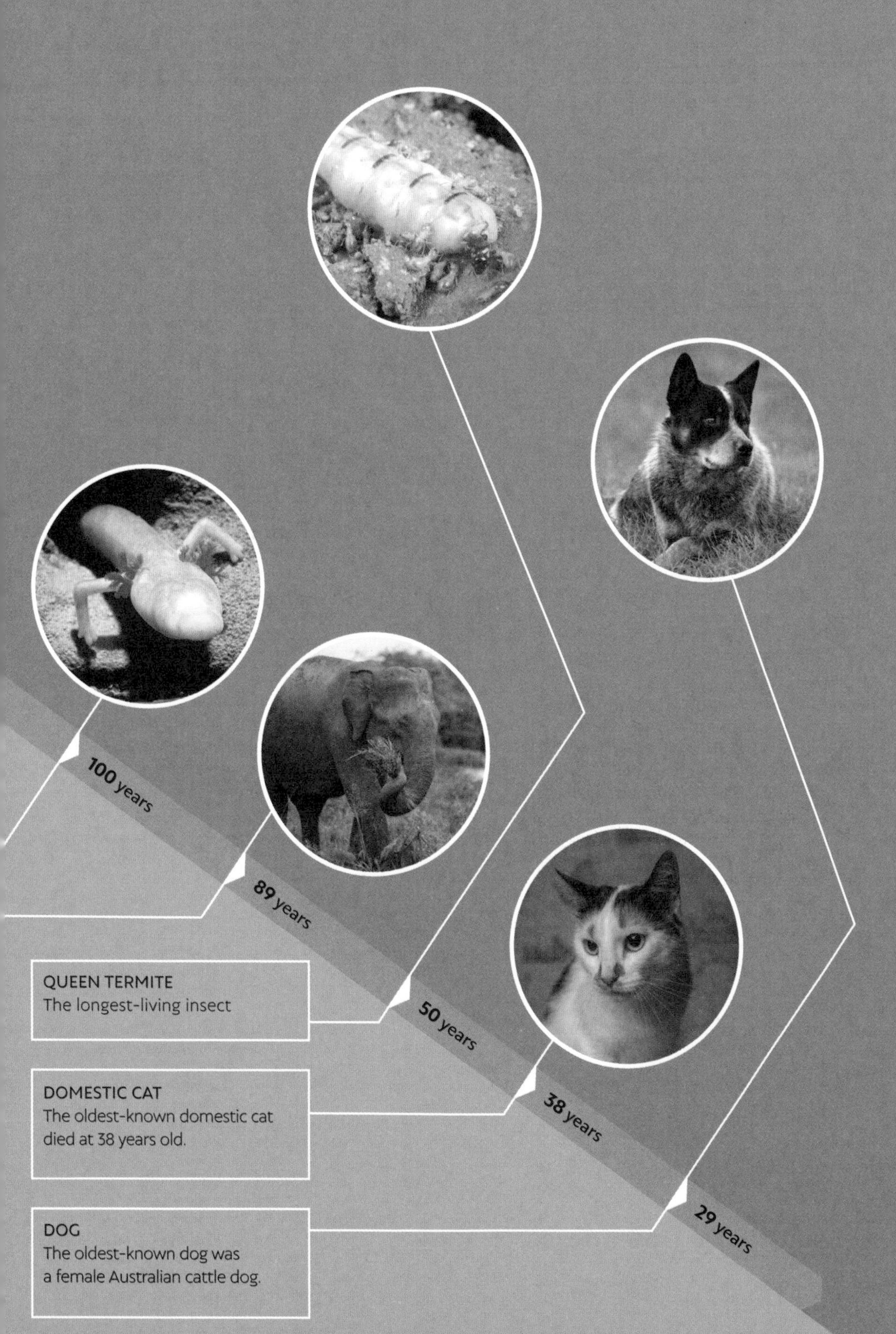

CLONAL COLONIES

Some species can reproduce asexually, meaning without sexual contact. As a result, their progeny are genetically identical—both to the original organism and to one another. These clones sometimes form massive colonies, with millions of individuals tightly packed into the same geographical area.

Some plant species form clonal colonies; an individual in the colony is referred to as a "ramet." When fungi form clonal colonies, they are connected via a common rootlike structure called a "mycelium." Although the individuals—we know them as "mushrooms"—may be relatively short-lived, the colonies they belong to can last for immense time spans. Indeed, they form some of the oldest-known organisms on Earth.

Global spread of the world's oldest clonal colonies

2,000–8,500 years old
Armillaria ostoyae located in Malheur National Forest, Oregon. This is a 3.4-square-mile (9 sq km) colony of mushrooms known as "The Humongous Fungus."

10,000–80,000 years old
Quaking aspen located in Fishlake Natural Forest, Utah. A clone called Pando (Latin for "I spread") is made of 47,000 male trees with a single, shared root system. Scientists who studied DNA from more than 200 individual tree's stems have confirmed that the colony is clonal. It is spread over 0.2 square miles (0.5 sq km) and weighs around 6,613 tons (6,000,000 kg).

13,000 years old
Jurupa Oak located in Jurupa Mountains, California. This colony of trees needs fire to grow. Its burned branches only sprout new shoots after they are burned.

12,000–100,000 years old
Neptune grass located in the Mediterranean Sea near Ibiza. Discovered in 2006, this 5-mile-wide (8 km) underwater meadow is thought to be one of the oldest clonal colonies on Earth. It is an important ecosystem, not least because a patch of Neptune grass can soak up 15 times more carbon dioxide every year than a similar-size piece of Amazon rain forest.

9,550 years old
Norway spruce located in Fulufjället National Park, Sweden. Old Tjikko is a single clonal tree that has regenerated new trunks, branches, and roots over millennia. Its current trunk is a few hundred years old. It stands 16 feet (5 m) tall.

3,000–13,000 years old
Mongarlowe mallee (critically endangered shrub) located in New South Wales, Australia. A type of eucalyptus, this smooth-barked shrub has white flowers and hemispherical fruit.

43,600–135,000 years old
King's Lomatia (critically endangered) located in Southwest National Park, Tasmania. Only one colony is known to exist in the wild. It contains about 600 individuals.

LIFE IN THE FAST LANE

They say life is short, and for many species, it really is. Their bucket lists had better be concise, because for many animals, their entire life is compressed into a few short months or weeks. Unsurprisingly, insects tend to have the shortest life spans, but some vertebrates have short lives, too. Müller's giant Sunda rat, for example, native to southeast Asia, is one of the shortest-living mammals, with an average life span of only 6 to 12 months.

With some animals, including insects and fish, it's easy to think that their lives are much shorter than they really are. The mayfly is well known for its fleeting 24-hour life span, but this is only the adult portion of its life. The mayfly's whole life is made up of different developmental stages and lasts for a year or more.

The life cycle of the mayfly

TOTAL LIFE SPAN: 1–2 years
DISTRIBUTION: Over 3,000 species worldwide

EGG STAGE

TIME SPAN: **A few days to a few weeks**
Eggs are laid on water. They sink to the bottom and stick to plants and stones.

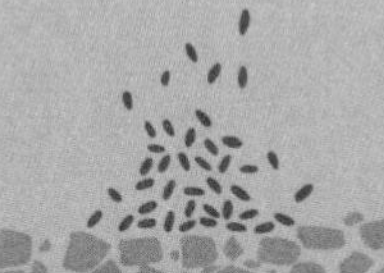

NYMPH STAGE

TIME SPAN: **A few months to a few years**
The larvae molt repeatedly as they grow. They feed on algae and detritus.

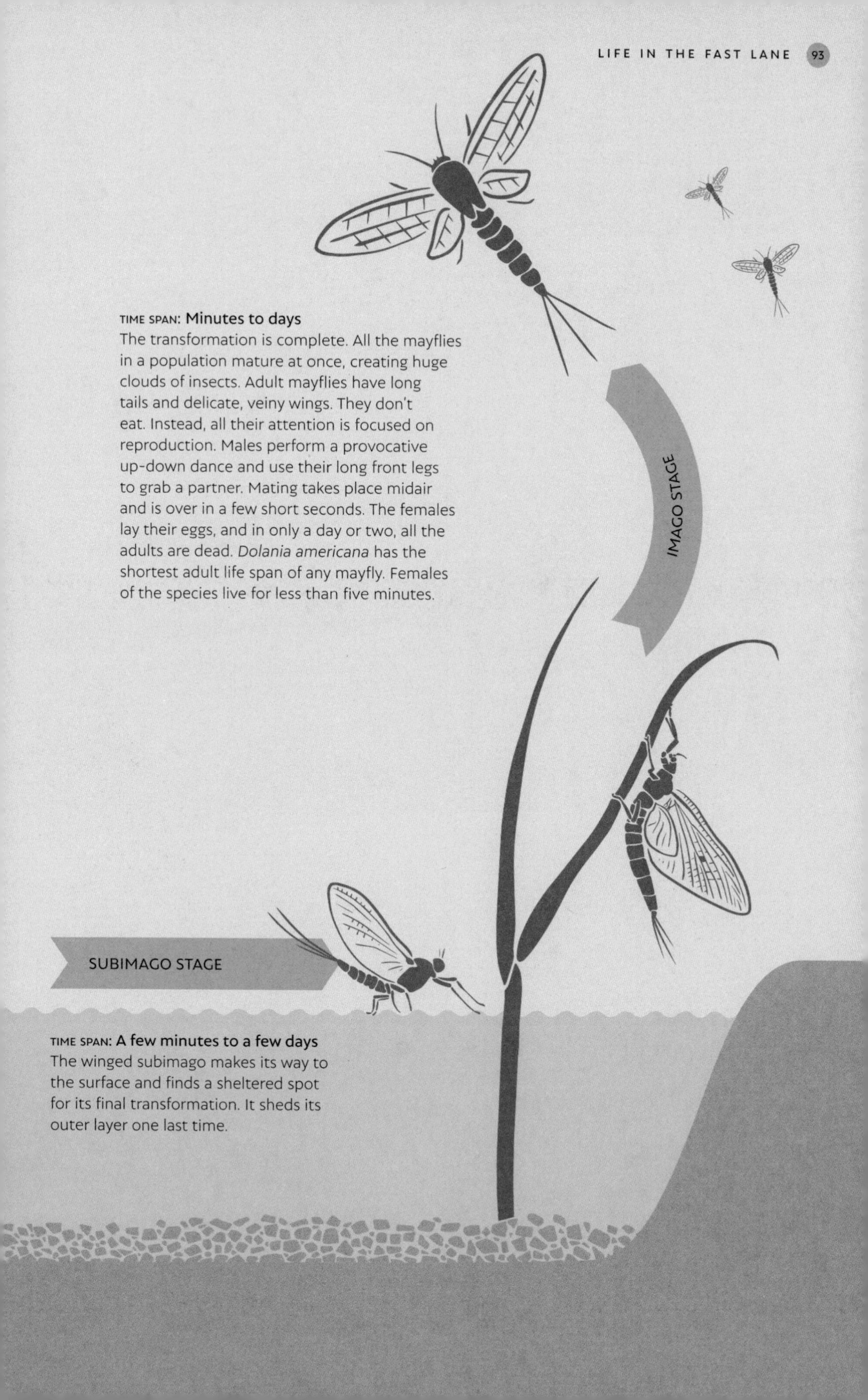

TIME SPAN: **Minutes to days**
The transformation is complete. All the mayflies in a population mature at once, creating huge clouds of insects. Adult mayflies have long tails and delicate, veiny wings. They don't eat. Instead, all their attention is focused on reproduction. Males perform a provocative up-down dance and use their long front legs to grab a partner. Mating takes place midair and is over in a few short seconds. The females lay their eggs, and in only a day or two, all the adults are dead. *Dolania americana* has the shortest adult life span of any mayfly. Females of the species live for less than five minutes.

TIME SPAN: **A few minutes to a few days**
The winged subimago makes its way to the surface and finds a sheltered spot for its final transformation. It sheds its outer layer one last time.

The life cycle of the honeybee

TOTAL LIFE SPAN: 6 weeks to 2 years
DISTRIBUTION: 11 species in Eurasia

Honeybees go through four different life stages: egg, larva, pupa, and adult. Their life stages and life spans vary in length, depending on their role in the hive and the time of year they hatch. Queens, for example, take 15 to 16 days to become adults. Workers take 21 days and drones take 24 days.

Adult queen bees typically live for around one to two years. Adult workers live for 15 to 38 days in the summer, 30 to 60 days in the autumn, and 150 to 200 days in the winter. It's thought these differences reflect both their workload and the amount of food that is available. Adult drones, in contrast, live for around 21 to 32 days in the spring and the summer, which are the only times they are produced.

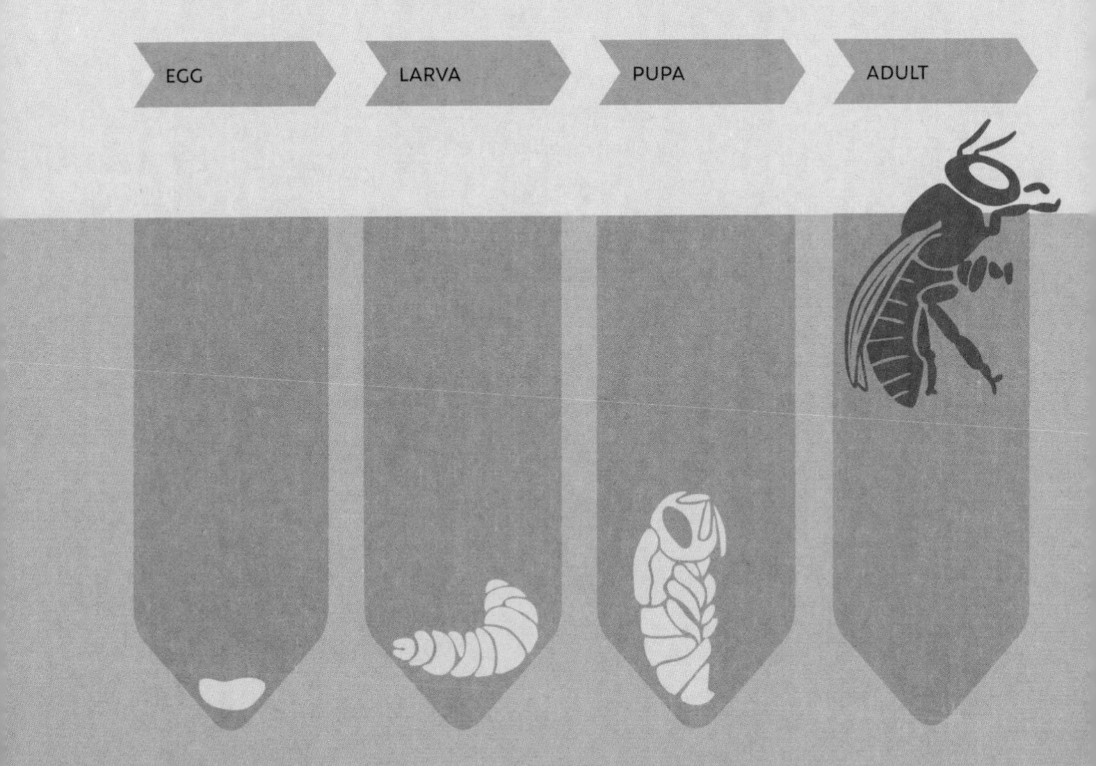

LABORD'S CHAMELEON

TOTAL LIFE SPAN: 1 year
DISTRIBUTION: Madagascar

This unusual animal spends more of its short, annual life inside the egg (eight months) than outside of it (four months). The eggs sit out the dry season and then hatch when the rains fall. By the time the new generation of chameleons hatch, in perfect synchrony with one another, their parents' generation is long gone. Juveniles reach sexual maturity in less than two months.

SEVEN-FIGURE PYGMY GOBY

TOTAL LIFE SPAN: 8 weeks
DISTRIBUTION: Indo-West Pacific

The seven-figure pygmy goby (also called the Sign Eviota) has the shortest-known life span of any vertebrate. It spends three weeks in the open ocean as a tiny larva before settling on a reef and taking one to two weeks to grow to maturity. Adults then live for three to four weeks. During that time, females can produce up to three clutches, each containing hundreds of eggs.

THE ELIXIR OF LIFE

If the ultimate life span is one that goes on indefinitely, without ever getting old, then a little jellyfish called *Turritopsis dohrnii* just might have found it. Also known as the "immortal jellyfish," this fingernail-size predator is found in warm waters all over the globe. When its adult form, or "medusa," gets old, damaged, or stressed, it reverts to an earlier stage in its life cycle and keeps growing. It's like pressing the rewind button on life, over and over again. This means that *Turritopsis dohrnii* can, in theory, live forever. In reality, however, they often get eaten.

The life cycle of *Turritopsis dohrnii*—the immortal jellyfish

REGULAR LIFE CYCLE

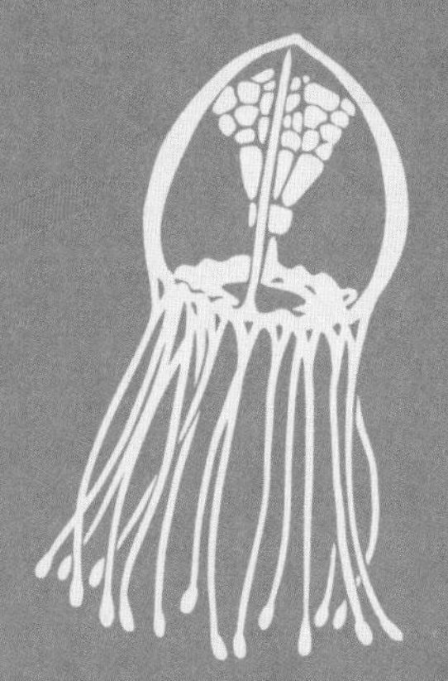

MATURE MEDUSA
It takes several weeks for the medusa to develop into a fully mature adult with 16 tentacles or more.

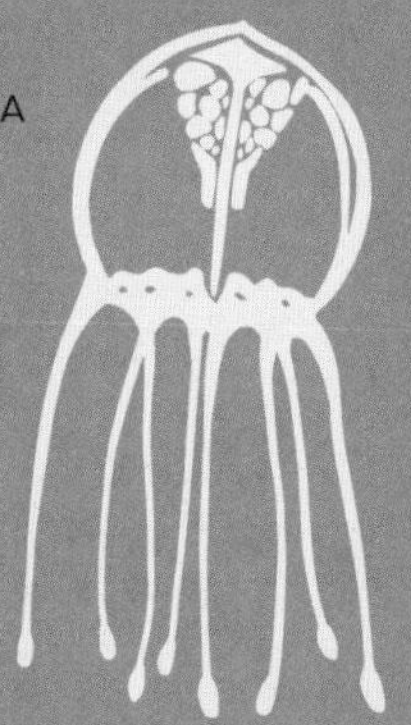

YOUNG MEDUSA
Adult jellyfish with eight tentacles

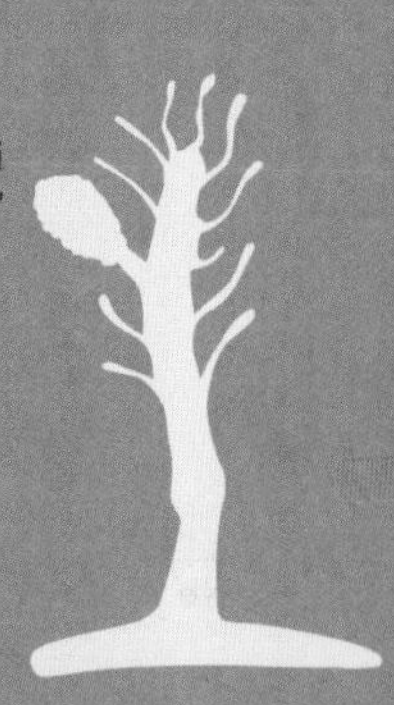

BUDS
The buds can form and break away to become mini adults known as "medusae."

POLYP
A small stalk topped with a tentacle-fringed mouth

DISCOVERY
The immortal jellyfish was first located in 1883 in the Mediterranean Sea, but it took 100 years for scientists to discover it had cracked the enigma of immortality.

Turritopsis dohrnii is not the only jellyfish able to cheat death. At least five other species of jellyfish are known to have regenerative abilities. The common jellyfish, found throughout the world's oceans, is also regenerative. Not only can the animal repair itself when body parts are damaged or destroyed, dying jellyfish can produce living polyps, which can live to see another day.

REPRODUCTION
Jellyfish can reproduce sexually and asexually. Polyps reproduce asexually by budding, while medusae spawn eggs and sperm to reproduce sexually.

HOW *TURRITOPSIS* CHEATS DEATH

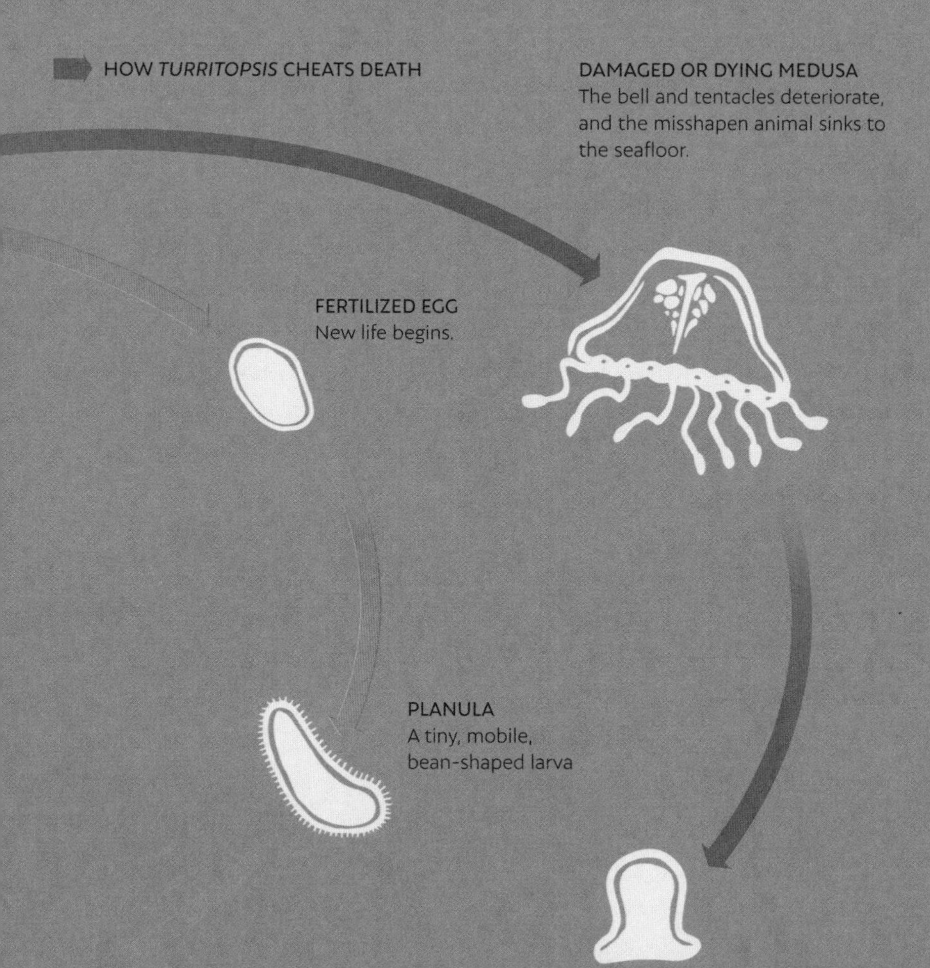

A LIFE SPAN LESS ORDINARY

Some life spans are remarkable, not for their length or lack of it, but for the dramas and transitions they contain. We are familiar with the eggs that turn into tadpoles and then become frogs, and with the exquisite life cycles of butterflies and moths, but the animal kingdom plays host to some truly bizarre life cycles. From the animal that is born pregnant to the caterpillar with its own personal bodyguard, sometimes the life spans of the natural world are packed full of surprises.

The life cycle of a freshwater pearl mussel

TOTAL LIFE SPAN: **150 years**
DISTRIBUTION: **Northern Hemisphere**

The freshwater pearl mussel is an endangered species of freshwater mussel. Its life cycle is unusual because it has both a regular and a parasitic stage, in which the mini mussels hitch a ride on a juvenile fish.

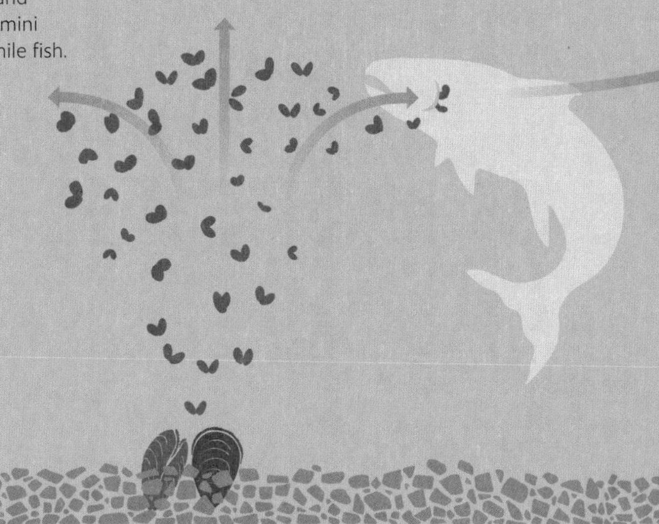

1–2 days

9 months

FERTILIZATION
Males release sperm into the water. Females "catch" what they can and fertilization occurs internally. The eggs develop into tiny larvae called "glochidia." They look like mini mussels.

RELEASE
Female mussels release millions of glochidia in a single, well-timed pulse. This usually occurs during July and September.

PARASITIC STAGE
Most of the glochidia are swept away, but a few are inhaled by juvenile trout or salmon and stick to their gills. The fish provide the developing mussel with fresh water, food, and a free ride to a new habitat farther upstream.

ECOLOGICAL HEROES

Freshwater mussels are ecological heroes. Through their filter-feeding activities, they purify the waters around them and help to keep our streams and rivers clean. Sadly, however, they can only take so much, and when the water quality declines, they struggle to survive. Over the past 100 years, there has been a 90 percent decline in populations in Europe.

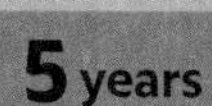

SETTLEMENT

In May or June the following year, the mini mussels drop from the fish and fall to the riverbed, where they burrow into the sand or gravel and remain hidden for years. They quietly continue to grow.

MATURATION

The tips of the mussels appear above the gravel bed. Adults grow extremely slowly. They are filter feeders, siphoning small organic particles, such as bacteria and algae, out of the water. They become sexually mature at 15 years old, and if left undisturbed, can live for over a century.

The life cycle of a chalkhill blue butterfly

TOTAL LIFE SPAN: 1 year
DISTRIBUTION: Parts of the Northern Hemisphere

The egg becomes the caterpillar, which turns into a chrysalis and transforms into a butterfly. We may be familiar with this classic story of metamorphosis, but what's notable about the chalkhill blue butterfly's life cycle is the company that it keeps. The insect spends about one-third of its life being protected by ant bodyguards.

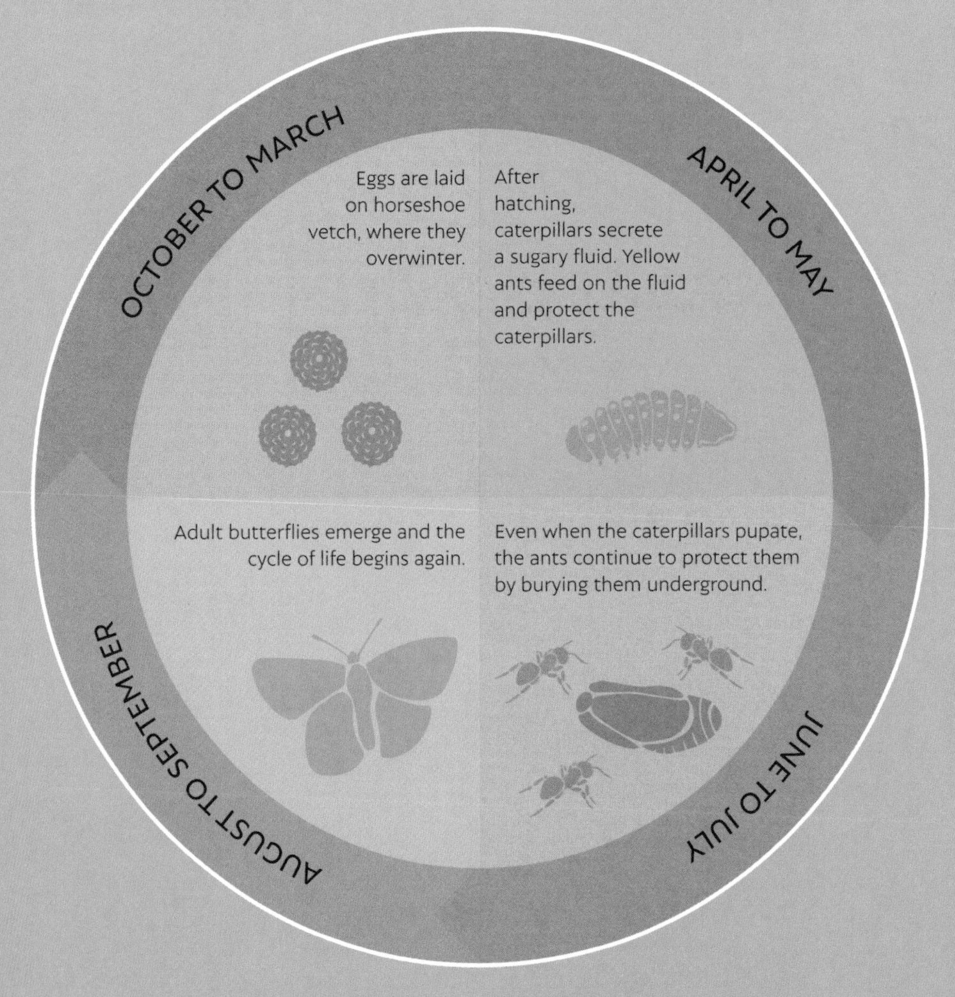

ADACTYLIDIUM MITE

TOTAL LIFE SPAN: Several hours to several days
DISTRIBUTION: Europe, Africa, North and South America

This tiny mite is like a warped Russian doll, because it is born with the next generation already inside it. Offspring are born pregnant. Each individual contains five to eight female mite larvae, plus a single male. This male mates with the females while still inside its mother. Then all of the larvae eat the mother from the inside out, and emerge pregnant, ready for the bizarre life cycle to start again. Newly emerged females go in search of food and die four days later, when they are eaten by their offspring. Newly emerged males die after a few hours.

CHALAZODES BUBBLE-NEST FROG

TOTAL LIFE SPAN: Uncertain
DISTRIBUTION: Southern India

Usually, frogs develop from tadpoles, which develop from eggs, but not always. Rare and secretive, the Chalazodes bubble-nest frog (*Raorchestes chalazodes*) was discovered in Kerala, southern India, in 1874. It then hopped under the radar for more than 100 years, only to be rediscovered at the turn of the new millennium.

Males and females of this critically endangered amphibian species mate away from streams and ponds, inside bamboo stems. The female lays five to eight eggs and then leaves the male to look after the eggs. They hatch, not as tadpoles, but as fully formed frogs.

IN THEIR PRIME

It's not uncommon for insects to have annual life cycles, but others have longer life spans. Most unusual, perhaps, are the periodical cicadas of North America, which synchronize their life span to fit with prime numbers: numbers that can only be divided by themselves and 1: for example, 2, 3, 5, 7, 11, 13, and 17. Periodical cicadas synchronize their emergence to a 13-year or 17-year cycle.

Why do these insects emerge on prime number–related cycles? One theory is that by running on 13- and 17-year cycles, the 2 different types of periodical cicada will only overlap once every 221 years. This means there is less competition for resources. Another theory, however, focuses on predators. Suppose there are some predators, such as the cicada killer wasp, that eat cicadas, and the cicadas emerge every 12 years. The predators that emerge every 2 years will kill them, as will those that emerge on 3-, 4-, or 6-year cycles. Emerging on a 13-year cycle would give the cicadas a better chance of survival.

The life cycle of the periodical cicada

TOTAL LIFE SPAN: 13 or 17 years
DISTRIBUTION: North America

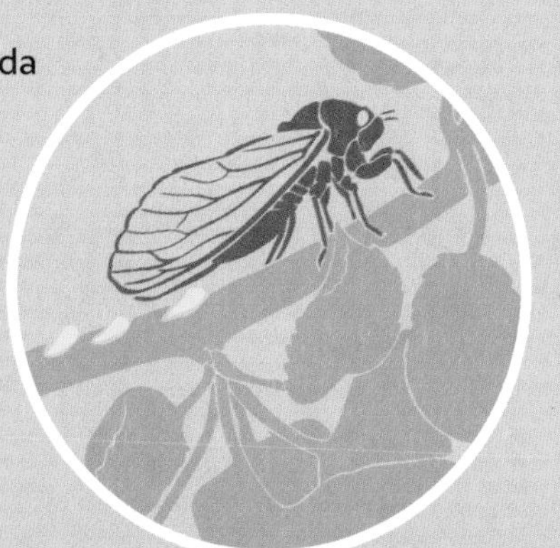

ADULT PHASE: 4–6 WEEKS
Huge numbers of cicadas emerge synchronously between April and June. Mating then takes place over a 10-day period, after which females lay hundreds of eggs in the stems of woody plants.

GOING UNDERGROUND

Periodical cicadas spend more than 99 percent of their lives underground. There are more than 3,000 species of cicada, but only 7 that are periodical.

LARVAL PHASE: 2–6 WEEKS
Inside the stem, the eggs hatch and turn into cylindrical larvae. The larvae then make their way down to the ground by dropping or crawling, then dig their way into the earth.

NYMPH STAGE: 13 OR 17 YEARS
The larvae develop into nymphs, which feed on juices of plant roots. As they grow, they shed their skin and bury deeper into the soil. In the year of their emergence, the nymphs excavate tunnels to the surface. This creates a little mound of mud, called a "cicada hut," before they finally emerge.

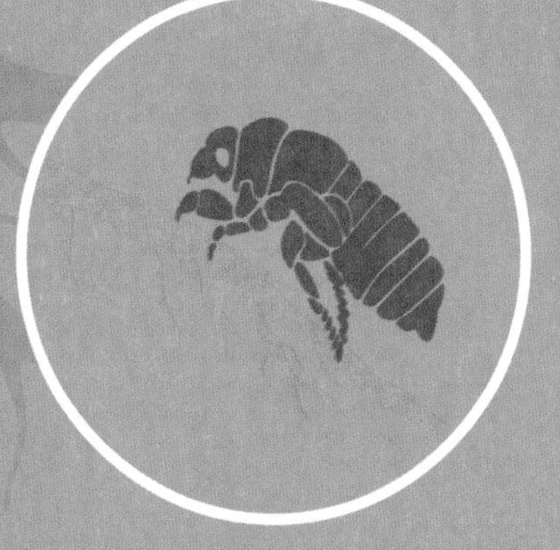

HIBERNATION

If the thought of snoozing away the winter months appeals to you, you are not alone. Many animals lower their heart rate, slow down their metabolism, and hibernate through a cold spell. Hibernation is a way to conserve energy when food supplies are scarce. Depending on the species and the surroundings, it can last for days, weeks, or months.

Hibernation periods

BIG BROWN BATS

DISTRIBUTION: **North America, Caribbean**

Big brown bats found in North America and the Caribbean hibernate alone or in small groups, in caves, mines, or warm human-made structures. In the wild, they hibernate for two months, but one captive big brown bat is known to have hibernated for 344 days—almost a whole year!

BROWN BEAR

DISTRIBUTION: **Eurasia and North America**

After eating lots of food in order to gain weight, most brown bears spend the winter hibernating in their dens. During this time, they recycle their proteins and urine, which helps them to avoid muscle wasting and disrupting their sleep. Adult females often give birth to their young inside the den and then spend the winter nursing them.

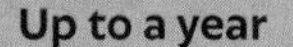

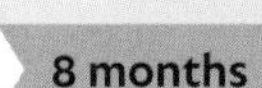

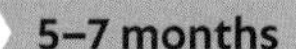

ALPINE MARMOT

DISTRIBUTION: **Europe**

As winter approaches, these large, ground-dwelling squirrels line their burrows with sticks and then seal themselves inside with a mixture of earth and their own feces. During hibernation, their heart rate drops to five beats per minute, and their breathing drops to a couple of breaths per minute.

Hibernation can also vary from long, deep spells of unconsciousness to shorter, lighter spells of inactivity interspersed with short periods of waking.

Hibernation sounds cozy but it can be dangerous. Hibernating animals can be vulnerable to predators, and they risk dying from a lack of body fat, severe weather conditions, or waking up too early to find that winter has not yet passed.

COMMON BOX TURTLE
DISTRIBUTION: US and Mexico

Turtles in the northern part of their range in the eastern US often hibernate over winter. They make themselves comfortable in the loose soil or mud at the bottom of streams and pools, or they slumber inside disused burrows.

COMMON POORWILL
DISTRIBUTION: North America

Many birds migrate to avoid the cold, but this nocturnal nightjar hunkers down in piles of stones and slumbers the winter away. It is the only bird known to hibernate.

5 months

Weeks to months

3 days

PYGMY SLOW LORIS
DISTRIBUTION: Southeast Asia

Native to Vietnam, Cambodia, Laos, and China, the pygmy slow loris also uses hibernation to conserve energy.

HIBERNATION IN THE WARM

Animals can also hibernate in warm weather, known as "estivation." Just like hibernating animals, estivating animals experience a lower metabolic rate and reduced physical activity. Some animals estivate to avoid extreme heat or drought, and the practice is common in warmer parts of the world. Many land-living and aquatic animals, including crocodiles, lemurs, and snails, estivate.

Torpor is another type of hibernation. During torpor, metabolic rate, body temperature, and physical activity are reduced. It helps animals to survive short periods of environmental hardship and typically lasts less than 24 hours. Hummingbirds, for example, undergo torpor during cold nights. Around 40 percent of Australia's land-living native mammals, including quolls, gliders, and kookaburras, also use torpor at some point in their lives.

Estivation periods

FAT-TAILED DWARF LEMUR
DISTRIBUTION: Madagascar

Hibernation of any kind is rare among primates. The fat-tailed dwarf lemur estivates during Madagascar's dry season. It uses the fat in its tail as a source of energy.

7 months

5 months

SALAMANDER FISH
DISTRIBUTION: Western Australia

This small fish from western Australia burrows into the mud to survive the drought period.

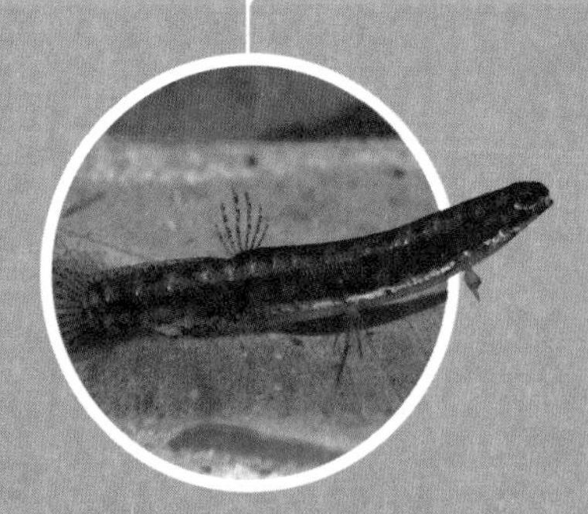

SNOOZING SNAILS

With their soft, delicate bodies, snails can dry out easily, which is why so many live inside shells. When it gets hot, some land snails estivate as a way to conserve energy and reduce the risk of desiccation. Some hide under detritus or dig their way underground, while others climb trees, plant stems, and fence posts to reach cooler layers of air. They secrete a membrane of dried mucus, called an "epiphragm," which they use to seal the opening of their shell. Some species reinforce this covering with a layer of tough calcium carbonate.

NILE CROCODILE
DISTRIBUTION: Africa

Crocodiles sit out the hot weather inside specially excavated burrows.

3–4 months

3 weeks

EARTHWORM
DISTRIBUTION: Worldwide

Earthworms avoid drying out by sealing themselves inside small chambers lined with their own mucus. They tie their bodies into knots to reduce the amount of surface area that is exposed to the soil.

BACK FROM THE DEAD?

Some organisms can be revived after huge periods of inactivity or dormancy. Microbes have been revived after millions of years. Tiny animals, such as nematode worms, have been revived after tens of thousands of years, and seeds can sometimes survive for millennia. Some microbes remain viable when they are trapped in amber or desiccated in salt crystals, while some animals

Animal dormancy periods

LENGTH OF FREEZE

32,000 years

30,000 years

24,000 years

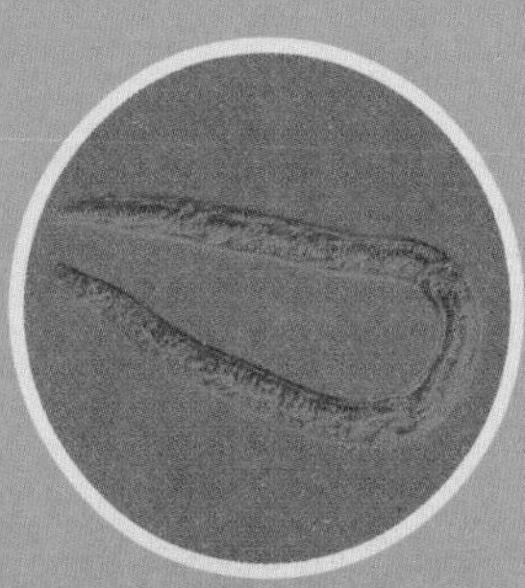

NEMATODE WORM
A couple of tiny nematode worms that were frozen in the permafrost for tens of thousands of years were brought back to life after scientists thawed and cultured them. For the short period of time that they lived, they were the oldest multicellular life forms known on Earth.

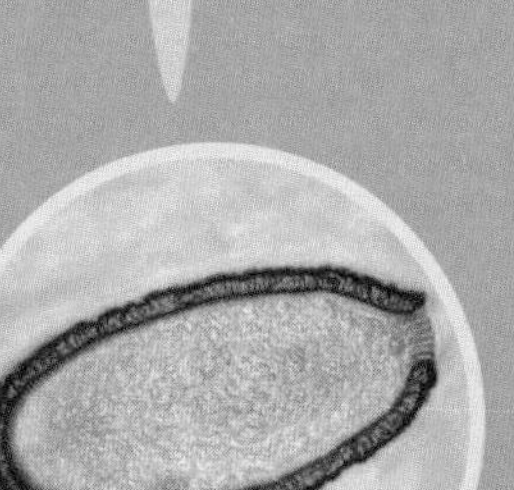

VIRUS
In 2014, researchers at the Genomic and Structural Information (GSI) laboratory revived an ancient virus from the Siberian permafrost by thawing it out and giving it something to infect. The virus, called *Pithovirus sibericum*, infects amoebae.

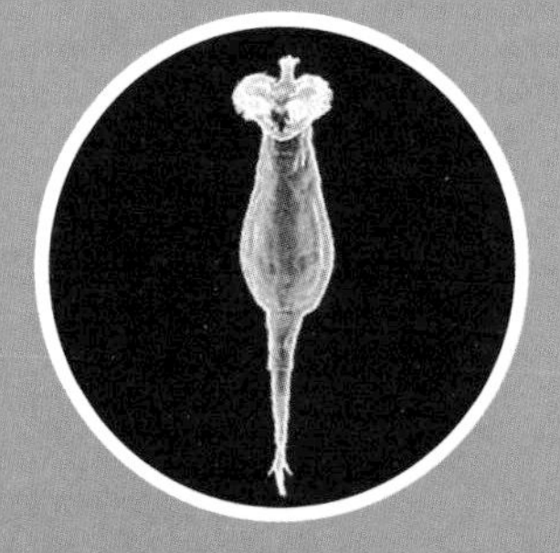

BDELLOID ROTIFER
This microscopic animal was revived after being frozen in the Arctic permafrost.

survive being frozen for remarkably long time spans. The trick, it seems, is to make your own antifreeze. The mountain stone weta, for example, makes an amino acid–rich cryoprotectant that helps its cells to stay intact, while up to 80 percent of this insect's body freezes and then thaws.

10 months

ARCTIC WOOLLY BEAR MOTH

The Arctic woolly bear moth spends seven years as a caterpillar, and it is frozen for most of this time. It can survive temperatures as low as −94°F (−70°C), and it only thaws for a couple of months per year, during which it eats, grows, and sheds its skin.

8 months

WOOD FROG

This little North American frog freezes and lies dormant for eight months each winter, then thaws in spring.

17 days

MOUNTAIN STONE WETA

This mouse-size insect, native to New Zealand's South Island, can survive temperatures of 14°F (−10°C) for two weeks or more, by allowing most of its body to freeze.

BACTERIA

Bacteria can remain dormant for 250 million years. Scientists collected, isolated, and then were able to grow samples of ancient bacteria harvested from salt deposits in New Mexico.

THE BEAL SEED EXPERIMENT

In 1879, American botanist William James Beal began an experiment to see how long seeds can remain viable. He buried 20 bottles containing sand and seeds from 21 common plants. At regular intervals, he exhumed the bottles, one at a time, and planted the seeds. After 10 years, most of the seeds were viable, but as time went on, fewer and fewer seeds germinated. Following his death in 1924, other scientists took over and the study continues to this day. The most recent bottle was dug up on April 15, 2021. Of the original 50 seeds in the bottle, 13 germinated, all from the same species: a flowering plant called the moth mullein. This shows that some seeds, at least, can remain viable for 142 years. Other seeds, however, can remain viable for much longer.

DOOMSDAY VAULT

The Svalbard Global Seed Vault, or "Doomsday Vault," in the frozen Far North, contains 642 million seeds. Most are seeds of food plants, including grains, fruits, and vegetables. With many plant species at risk from climate change and other threats, this provides a crucial reserve of plant species.

Length of dormancy in plants

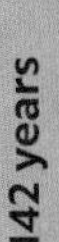

MOTH MULLEIN

The moth mullein is a flowering plant native to Eurasia and North Africa. Its seeds can stay viable for more than 100 years.

620 years

CANNA LILY

A seed placed inside a nutshell, which formed part of a necklace retrieved from an ancient Argentinian tomb, is the oldest viable seed known to date.

1,200 years

SACRED LOTUS

Seeds from 200 to 1,200 years old were recovered from a dry lake bed in China. Many of them germinated and went on to produce seedlings.

2,000 years

JUDEAN DATE PALM

Seeds from this previously extinct tree were found in an ancient jar at Herod the Great's Palace on Masada, Israel, and planted out. One of them, called Methusaleh, is now over 11 feet (3.5 m) tall and has started producing pollen.

32,000 years

NARROW-LEAFED CAMPION

In 2007, 600,000 seeds were recovered from ancient squirrel burrows, frozen deep in the Siberian permafrost. Most of the seeds were damaged, but scientists were able to isolate reproductive tissue from three immature seeds and use it to grow new plants. The plants flowered and produced viable seeds of their own.

4

GROWTH SPANS

INTRODUCTION

Growth is a fundamental characteristic of life fueled by cells, which divide to make more cells, forming the tissues, organs, and systems that make up all living things.

Typically, growth tends to be fastest during the early stages of life. Relative to their small size, for example, seedlings grow with alacrity. Embryonic animals, too, are a maelstrom of cell division and rapid growth, while youngsters and adolescents require enormous amounts of food to power their continued development into adulthood. In the middle and later stages of life, growth rates decline and often grind to a halt, but not for all animals. In captivity, the only thing that seems to limit the growth of a Mexican axolotl is the size of its tank. These unusual amphibians remain in their larval stage throughout life and keeping growing well into adulthood.

There are, of course, enormous differences in the growth spans and rates that exist between species. The African killifish grows rapidly and becomes sexually mature within 17 days, yet in colder waters, the Greenland shark grows seemingly at leisure. An individual can take up to 150 years to reach sexual maturity and can live on for several centuries.

By and large, growth follows a predictable trajectory. A female horse, for example, gives birth to a foal after a year of pregnancy, after which the little animal takes time to grow to maturity. If it loses a leg or an organ becomes damaged, things do not bode well. Mammals such as horses and humans possess limited regenerative powers. Other animals, such as axolotls, starfish, and flatworms, can spectacularly regrow parts of their bodies. It takes weeks to months, and in some instances, tiny fragments of body can morph into entirely new organisms. Some starfish can regenerate their whole body from just a single arm, while flatworms can be sliced, diced, and separated into tiny pieces, each of which can regrow to form a new adult. Growth is a truly remarkable process.

Starfish have amazing regenerative abilities. They can regrow legs and larger parts of the body.

THE GREAT EGG RACE

They can be hard and brittle, or soft and jellylike. Eggs are the membrane-bound structures that contain the developing embryos of animals such as birds, amphibians, and fish. Often, parents protect their eggs by incubating, or "brooding," them. In some species, such as the Atlantic canary and the giant Pacific octopus, the female does all the incubation, but in others, such as the cassowary and the common coqui frog, only the male incubates. The duties can also be shared or ignored entirely. Incubation spans can range from days to years, and hatching time can also be influenced by environmental factors, such as temperature and humidity. In many species of reptile, temperature also plays an additional role by determining the sex of the offspring. Red-eared terrapins, for example, all emerge from their eggs after three months or more, but eggs that are kept at 72°F–81°F (22°C–27°C) emerge as males, while those incubated at warmer temperatures emerge as females.

How many days to hatch?

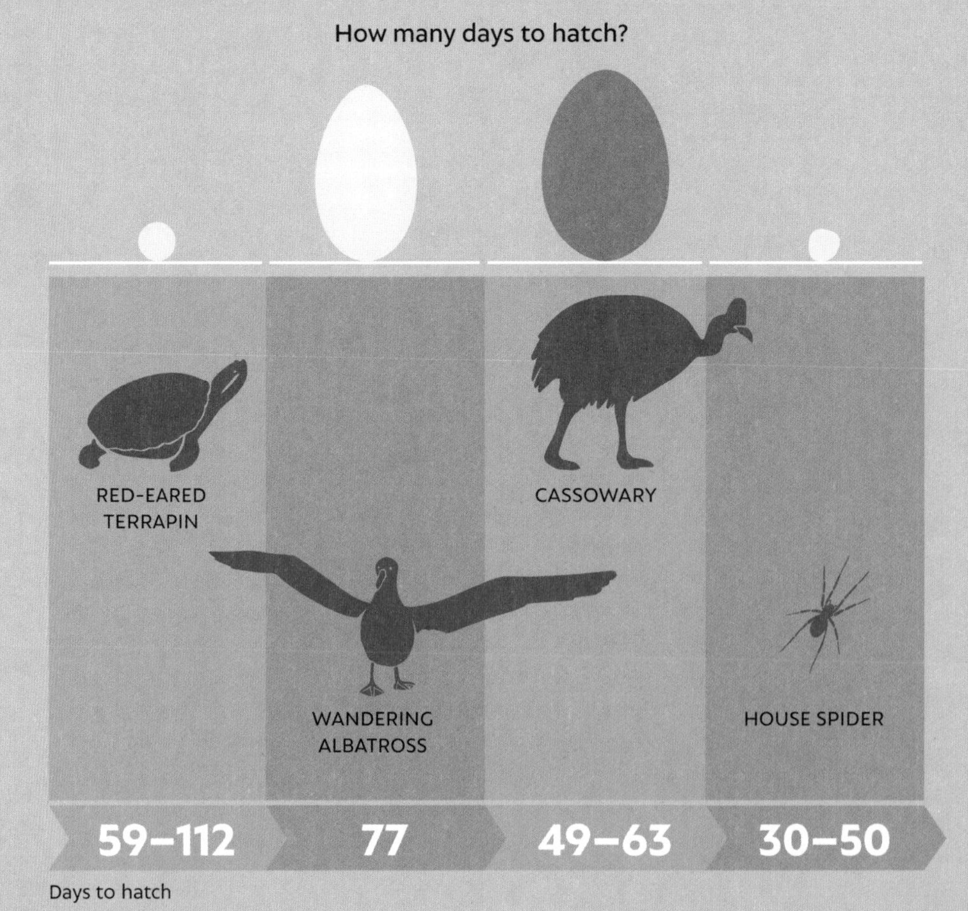

Days to hatch

ONCE IN A LIFETIME

Octopi are semelparous, meaning they only have babies once in their lifetime. They can lay tens of thousands of eggs over several days, which the female cares for. She cleans them, wafts currents of oxygen-rich water over them, and chases away predators. The female giant Pacific octopus, for example, broods her eggs for around 5 months, but in colder waters, such as those off Alaska, it may take up to 10 months for the eggs to hatch. One species of octopus, *Graneledone boreopacifica*, has the longest-known incubation period of any animal. Researchers spied on one ghostly white female, in a deep canyon in California's Monterey Bay, via a submersible that paid repeated visits. The mother brooded her young for 53 months. During this time, she didn't eat or leave her eggs. She died from exhaustion shortly after they hatched.

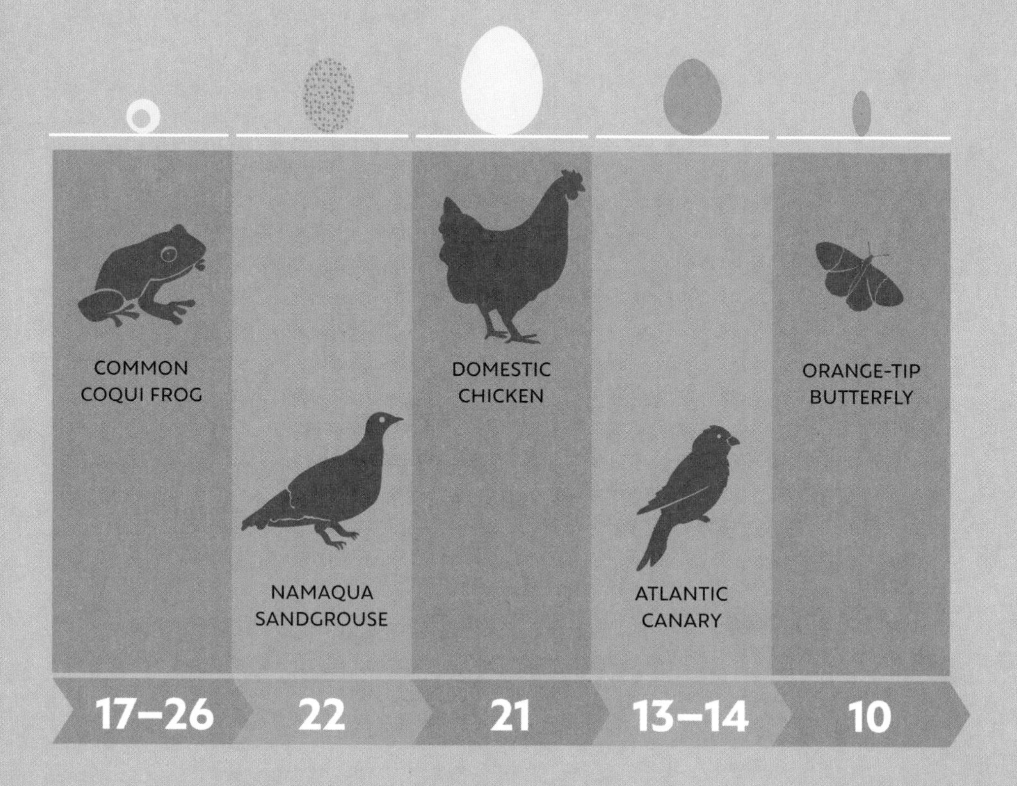

WHICH CAME FIRST?

Which came first, the chicken or the egg? The answer depends on who you ask. An evolutionary biologist will tell you that animals have been laying eggs for millions of years. Some of the oldest-known fossilized amniote eggs are from a dinosaur called *Massospondylus* that lived around 190 million years ago. These are the type of egg in which the developing embryo is surrounded by a series of membranes, like in a modern chicken's egg. The first recognizable domestic chicken, however, did not strut its stuff until 4,000 years ago. Therefore, eggs clearly came first.

Embryonic development in the chicken

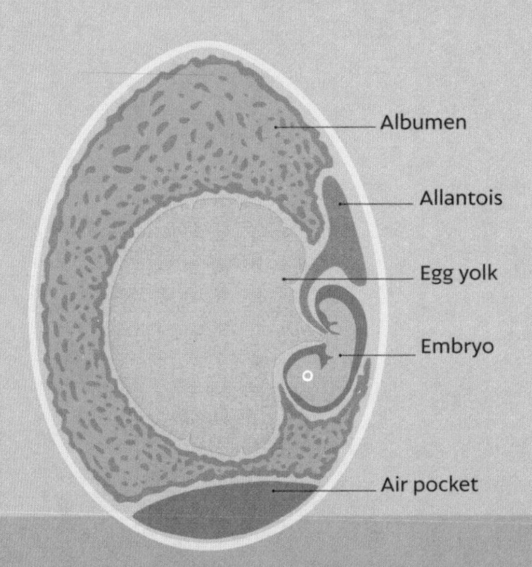

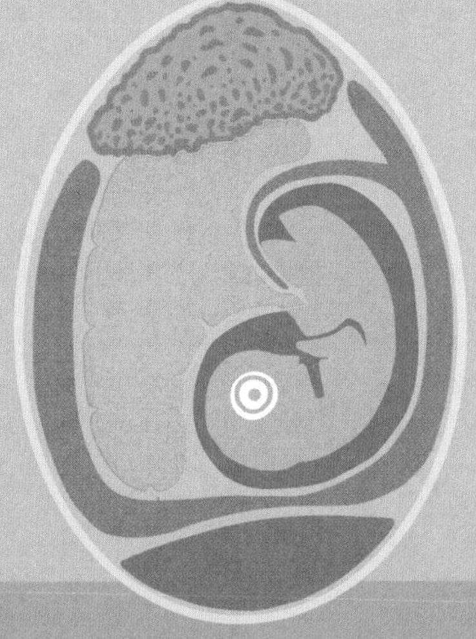

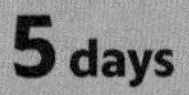

Beak, limb buds, and wings are developing. Internal organs have formed. The heart is beating.

10 days

The embryo is growing rapidly. Feathers are starting to form. The wings contain digits.

ALBUMEN
The albumen, or egg white, protects the developing chicken and shrinks as the embryo grows.

ALLANTOIS
The allantois stores waste and enables respiration.

EGG YOLK
The yolk provides food for the developing chicken and shrinks as it is used up.

Ask a developmental biologist, however, and they will approach the problem from a different angle. Think about how a chicken develops. After fertilization, the developing chicken is briefly a single cell, which then starts to divide. Inside its mother, the adult female's reproductive system is like an automated assembly line. As the dividing ball of cells, which contains the yolk and a tiny spot of DNA, begins to move along the assembly line or oviduct, the white and the shell are added. The whole process takes 24 hours, and the embryo-containing egg is laid by the bird and incubated. So, the chicken came first, and was followed by the eggshell that surrounds it.

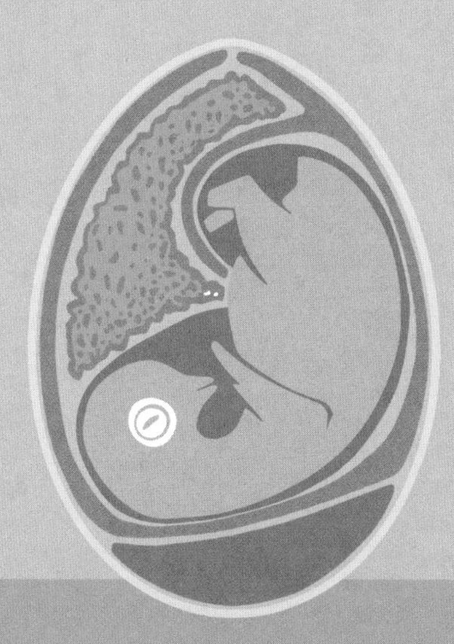

15 days

The embryo is increasingly birdlike. Scales, claws, and beaks harden up.

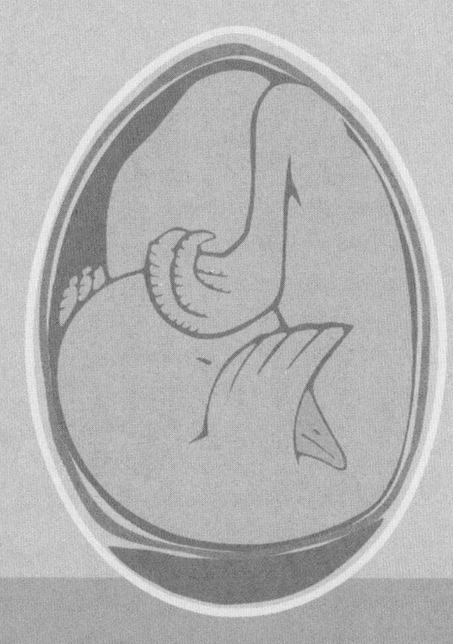

20 days

The embryo fills almost the entire egg. It pierces the air cell with its beak and takes it first breath.

21 days

Hatching takes 4–12 hours.

PREGNANCY IN MAMMALS

Several key factors influence the length of a mammal's pregnancy or "gestation period." Larger animals tend to be pregnant for longer, as do those whose offspring are highly developed at birth.

Elephants have the longest pregnancy of any mammal, lasting 650 days. Their young walk within two hours and are born with advanced cognitive skills. Species with highly developed newborns are called "precocial." Rats, in contrast, are altricial. Their babies are born blind and helpless. It takes three weeks before they are weaned and able to survive on their own.

Scientists have monitored the way embryonic elephants develop, using ultrasound technology. Pregnancy can be identified at day 50, with the first heartbeats visible from around day 80. The eyes are visible by day 97, the trunk by day 104, and the big ears 60 days later. The information is useful for zoos, where pregnant elephants often receive ultrasound checkups.

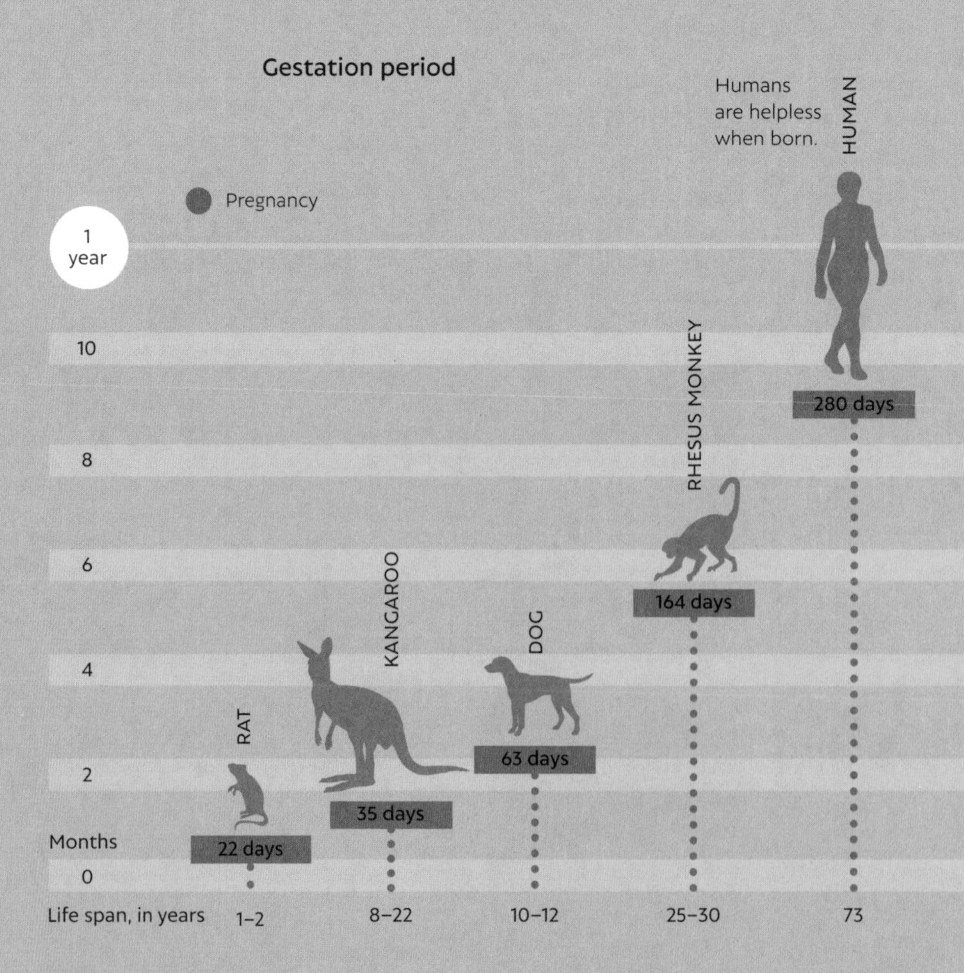

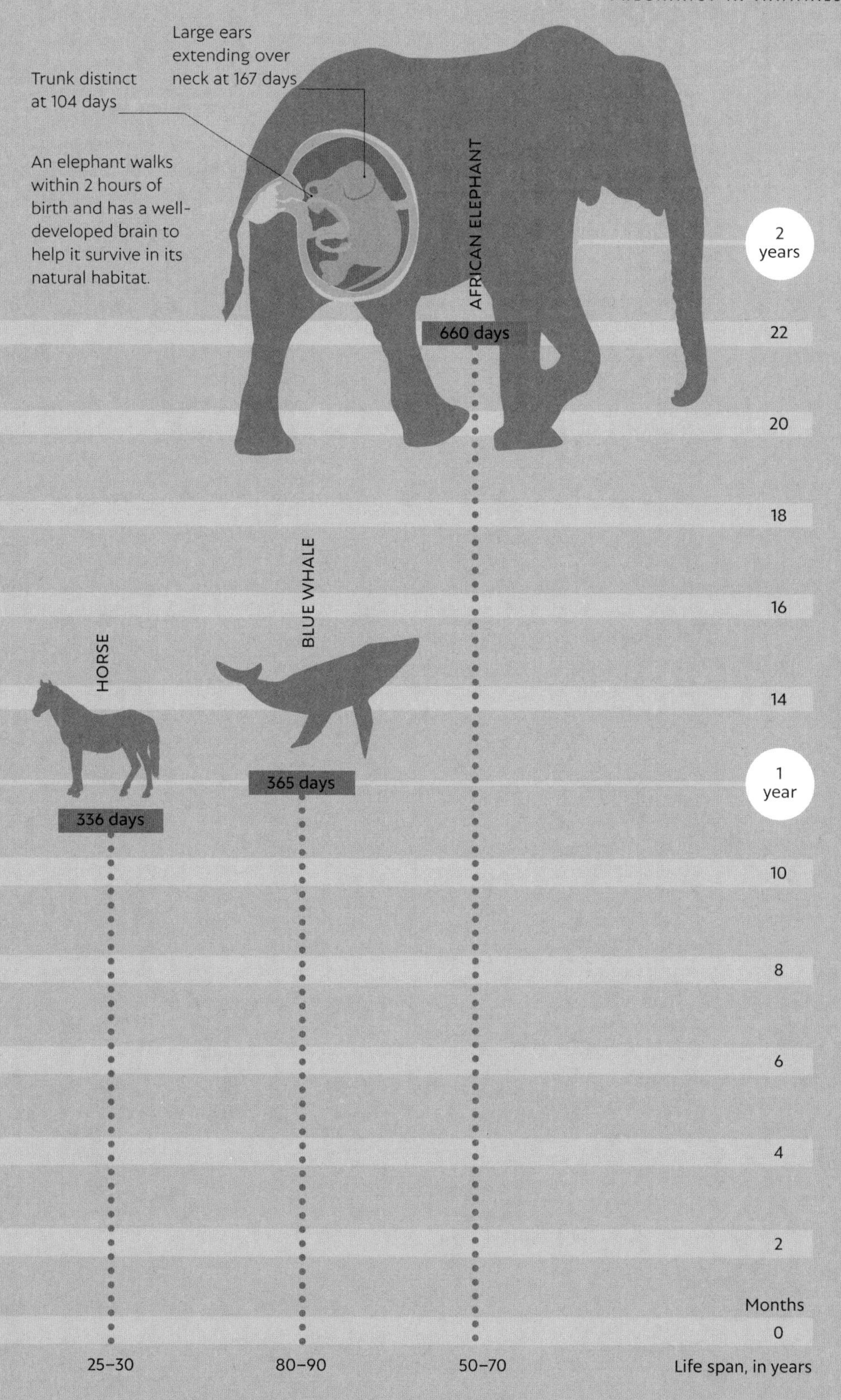
Large ears extending over neck at 167 days
Trunk distinct at 104 days
An elephant walks within 2 hours of birth and has a well-developed brain to help it survive in its natural habitat.
AFRICAN ELEPHANT
660 days
BLUE WHALE
365 days
HORSE
336 days
2 years
22
20
18
16
14
1 year
10
8
6
4
2
Months
0
25–30
80–90
50–70
Life span, in years

Guinea pigs are also precocial. Mothers have a relatively long pregnancy, of around 65 days, and then give birth to babies that are like mini adults. They have open eyes, fur, teeth, and claws, and they can eat solid food almost immediately. Precociality is a useful strategy in species where the young are not easily protected. Hares, for example, make their homes in open fields, so their young (leverets) are born precocial.

SAFEGUARDING THE UNBORN

Some animals have evolved clever strategies to protect their unborn. In North America, for example, little brown bats mate in the autumn. Females store the sperm internally through the cold winter months and then fertilize their eggs and give birth in the spring. This is called "seasonal delayed implantation."

Fertilization occurs immediately after the nine-banded armadillo has sex, but the development of the embryo is put on hold for three to four months. After this period of "embryonic diapause," the embryo then implants into the womb and splits into four identical quadruplets, which are born five months later. If the armadillos feel stress, which is often caused by living in captivity, diapause can be extended by one year or more.

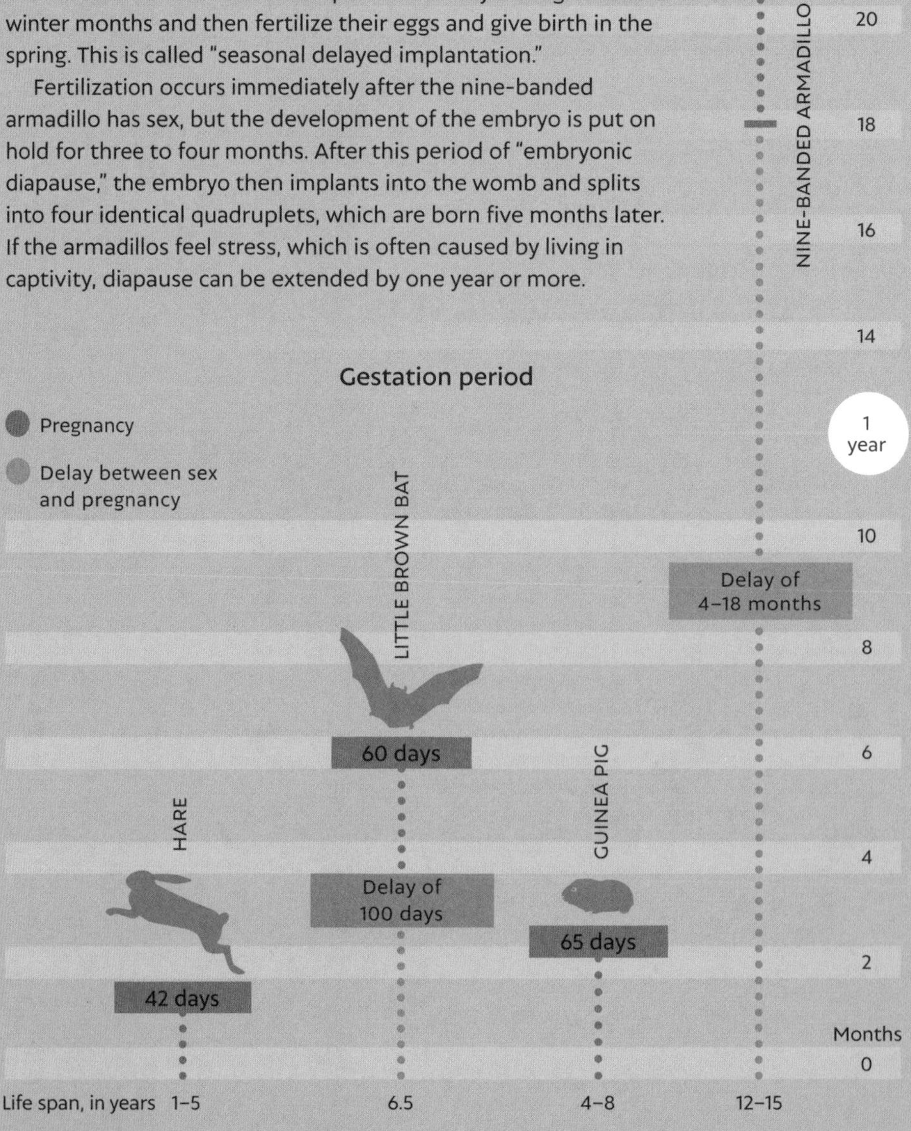

PREGNANCY IN MARSUPIALS

Most marsupials are large mammals yet have short gestation periods. Kangaroo pregnancies last only 35 days, after which up to four baked bean–size babies crawl into their mother's pouch. They stay there for 11 months, suckling the energy-dense milk that is released via the mother's four nipples. Although the mother may mate again during this time, the next offspring are not born until the pouch has been permanently vacated. The female kangaroo simply suspends the pregnancy until the time is right.

KANGAROO
Pregnancy
35 days

OPOSSUM
Pregnancy
12 days

The North American opossum is pregnant for only 12 days and gives birth to 16–20 jelly bean–size young. Newborns crawl straight from the birth canal to the pouch, but it is a dangerous journey and fewer than half of them survive.

PREGNANCY IN MONOTREMES

The duck-billed platypus and the echidna, both from eastern Australia, are egg-laying mammals known as "monotremes." Twenty-eight days after mating, female platypi lay up to three leathery, soft-shelled eggs in a specially built burrow, then incubate them for around 10 days. On hatching, the young are blind and hairless. They feed on their mother's milk, which is released through pores in her skin.

Four species of echidna (spiny anteater) also lay eggs. The female lays just one egg, 22 days after mating, and immediately puts it into her pouch. The young echidna, called a "puggle," hatches 10 days later and suckles from a milk patch within the pouch, where it stays for up to 55 days.

PLATYPUS
Pregnancy
28 days

GROWING UP

The childhood, or "juvenile," period is one of the most important phases in an animal's life. It is when they grow and mature, and learn and practice the skills that are needed to see them through adulthood. Some juveniles look very different to their adult counterparts. Baby Brazilian tapirs, for example, have a brown coat with white stripes and spots. The markings act as camouflage, but they disappear as the infant grows, maturing into adults that are brown all over. Appearance, however, is not always a reliable way to judge if adulthood has been reached. Some humans, for example, continue to act immaturely well beyond their teens, so biologists use other measures, such as generation time or the time taken to reach sexual maturity.

BRAZILIAN TAPIR WITH YOUNG

Time to reach sexual maturity

150 years
GREENLAND SHARK
The Greenland shark, which can live for several centuries, is an apex predator. With few natural threats, it takes its time to reach adulthood. It has the longest known juvenile period of any animal.

10–20 years
TUATARAS
After hatching from soft, parchment-like eggs, these reptiles, which are native to New Zealand, mature slowly. They become fertile after a decade but continue to grow in size for up to 35 years.

7–8 years

GIBBON

These tailless apes live in the rain forests of southern Asia. After reaching sexual maturity, gibbons often pair for life, but this does not stop them from having the odd fling with other gibbons from time to time.

7 months

COMMON MOUNTAIN VISCACHA

Age is relative. The common mountain viscacha is a rabbitlike rodent that lives in the Peruvian Andes. It becomes sexually mature at seven months old but dies one or two years later. It is the equivalent of a human exiting adolescence in their 30s.

3–5 months

EUROPEAN RABBIT

Bucks reach sexual maturity at four months, while does begin breeding between three to four months.

17 days

AFRICAN KILLIFISH

In some species, where life is precarious and predation rates are high, evolution favors the development of rapid growth and sexual maturity. This enables individuals to reproduce and pass their genes on before their lives are abruptly terminated.

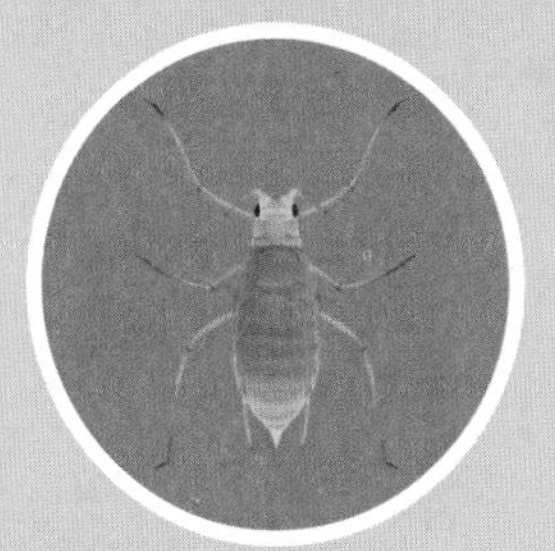

4 days

APHID

The aphid *Rhopalosiphum prunifolia* has the shortest-known generation time of any animal. It takes four days for one generation to replace its predecessor, although this time span is heavily influenced by environmental factors such as temperature and food availability.

THE AMPHIBIAN THAT NEVER GROWS UP

Just like Peter Pan, some animals don't ever grow up. The axolotl is a type of salamander, native to just one remaining freshwater lake in Mexico, where it is critically endangered. Most amphibians morph from a water-dwelling juvenile to a land-living adult, but the axolotl remains aquatic even in adulthood. Axolotls remain in the water in their juvenile form through life. Instead of undergoing metamorphosis like most amphibians, they retain their three pairs of feathery gills, which help them to breathe in the water, and a distinctive caudal fin, which runs along the animal's back and helps it to steer. They reproduce via paedogenesis, which is the ability to reproduce in a preadult or larval form.

Although adult axolotls remain in their juvenile form, they still retain the ability to morph into a different adult version. When captive axolotls are treated with iodine, it triggers the production of thyroid hormones, which prompts artificial metamorphosis. Gills and fins disappear, eyelids develop, and limbs become more muscular. The skin becomes less permeable to water, and the lungs, which are already there, develop further. Altogether, the changes prepare the axolotl for a new life on land, but these changes never occur naturally. Captive axolotls treated in this way often die, so the procedure is not recommended.

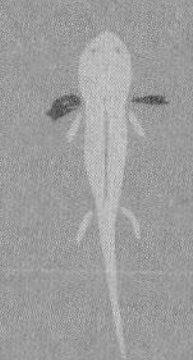

HATCHING
Inside the egg, axolotl embryos take two weeks to develop into larvae. They hatch after two weeks.

3 WEEKS
Freshly hatched axolotls emerge able to swim. Their legs are yet to develop.

3 MONTHS
The young juvenile continues to develop. It grows lungs and front and back limbs.

Extended juvenility in the Mexican axolotol

Axolotls remain in their juvenile form throughout life and continue to grow. They can live for up to 25 years in captivity, but in the wild their life span is usually considerably shorter.

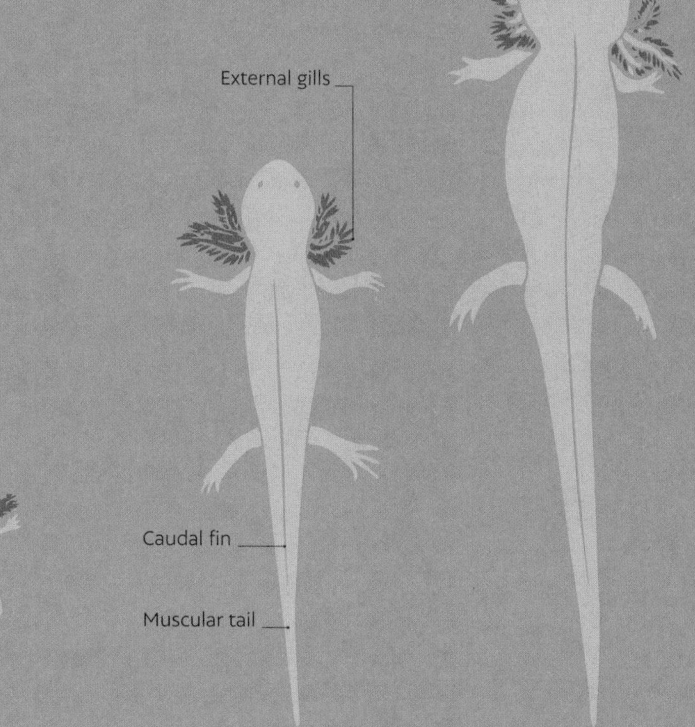

Life span up to 25 years

5 MONTHS
The late juvenile continues to grow rapidly, but it has yet to reach sexual maturity.

1 YEAR
Females become sexually mature at one year old, while males achieve this feat at nine months old.

3 YEARS
Although they remain in their juvenile form and continue to grow, age-related changes do occur. For example, the skin thickens and the skeleton becomes bonier.

ON YOUR FEET

Human babies usually begin to crawl between 6 and 12 months old. Toward the end of this time, they start to pull themselves up and then "furniture glide" around the room, grabbing on to any object within reach. Between 8 and 18 months, they take their first proper steps, followed almost immediately by their first proper face-plant. In the weeks that ensue, the art of walking is perfected.

How long does it take baby animals to walk?

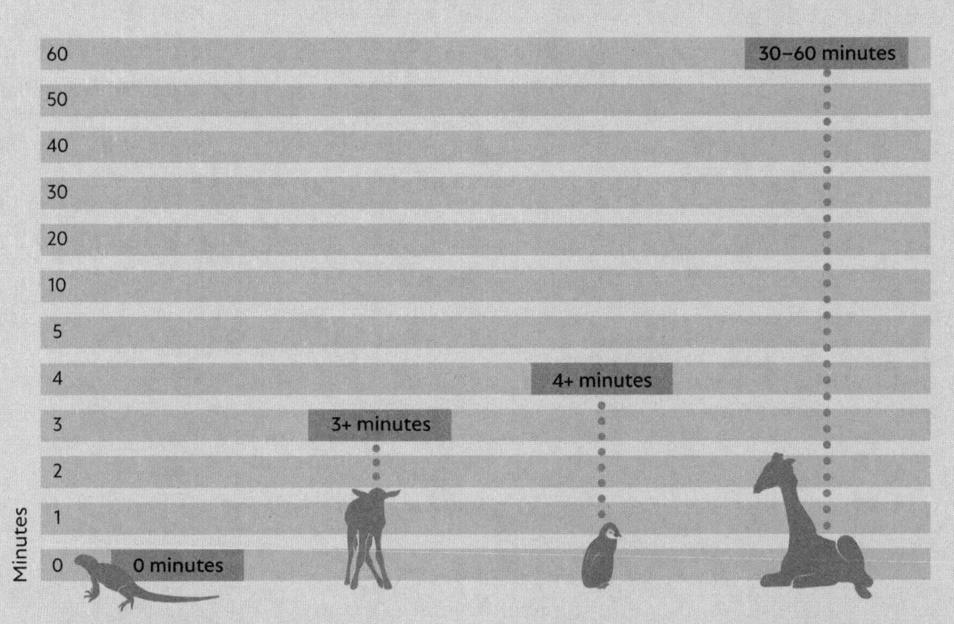

MARINE IGUANA

Marine iguanas, which are native to the Galápagos Islands, can run as soon as they hatch. Surrounded by predatory snakes, the freshly emerged reptiles must race from their land-based hatching spot to the relative safety of the ocean where they can graze on algae.

GNU

Hundreds of thousands of baby gnu are born at the start of the African rainy season, in February and March. They can walk within minutes and a few days later begin to trot alongside their mothers as part of the rapidly growing herd.

PENGUIN

If you're a flightless bird, being able to walk is extra important. Penguin chicks can stand almost instantly and waddle not long after. Unlike humans, who divide their weight between both feet when walking, penguins place all their weight on the front foot. This helps them to walk on the ice without slipping.

GIRAFFE

Baby giraffes are born almost 7 feet (2 m) tall. They quickly manage a wonky walk, but it takes several hours to stand up straight and walk with style. Within 10 hours, they can run at full speed and keep up with the adults in their families.

When it comes to walking, we humans are outliers. We spend far longer learning to walk than most other animal species, largely because we grow up in such a pampered and predator-free environment. As a rule of thumb, however, predator species take to their feet more slowly than those species that are prey. It makes sense. If a vulnerable species is born into a world surrounded by hungry assailants, being fast footed or "hooved" helps them to dodge danger and boosts their chances of survival.

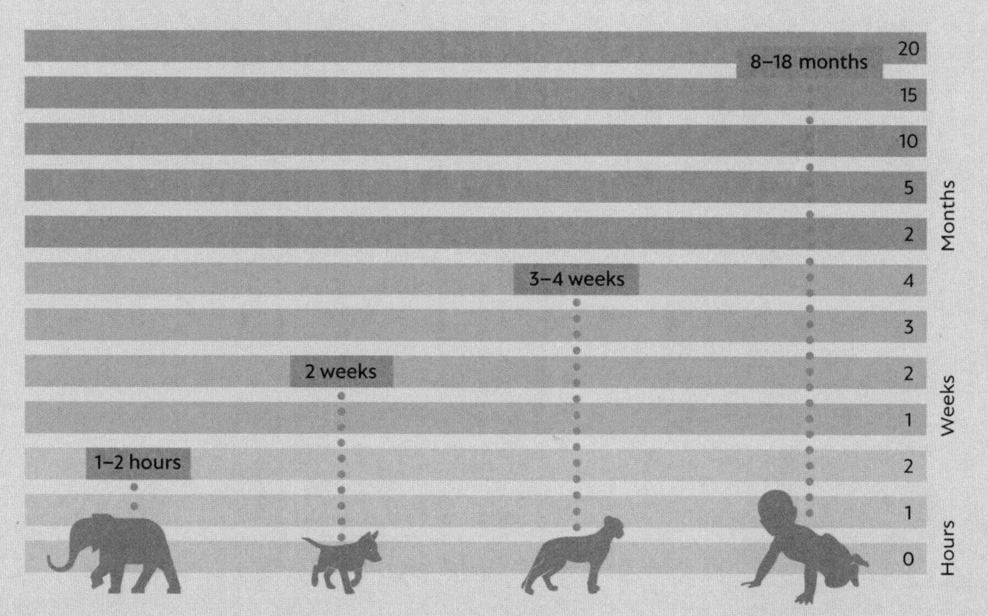

ELEPHANT
Elephant herds need to keep moving in order to find food and water, and there's no letup when a baby is born. A 2022 study by scientists at Oxford University, UK, found that, although the speed of the herd may dip a little on the day of birth, it quickly picks up, and even day-old babies are able to keep up.

WOLF
Four to six pups are born inside the den. They are blind, deaf, and largely helpless, but they grow fast. By two weeks of age, their eyes are open and they are tottering about. They leave the den for the first time one week later.

CHEETAH
Just like humans, cheetah cubs need care after they are born. They can walk at a few weeks old and run shortly after. From six months, cheetahs receive hunting lessons from their mum. They strike out on their own about 18 months later.

HUMAN
Humans take longer than other animals to walk because of the way that our brains develop. Big-brained babies would struggle to pass through the relatively narrow birth canal, so we are born with immature brains that continue to develop after birth. We acquire the ability to walk only when our brains "catch up" with our bodies.

REEF BUILDING

Although they don't look like a typical animal, corals are part of the animal kingdom. They belong to the group Cnidaria, which also includes sea anemones and jellyfish. These ecosystem engineers live in vast underwater colonies. Each individual coral is made up of thousands of much smaller, genetically identical organisms called "polyps." Each polyp comprises a tubelike stomach topped with a tentacle-fringed mouth. There are thousands of types of coral, but the polyps of reef-building coral secrete a rigid skeleton of calcium carbonate, which cements them to their neighbors and builds up over time to create the reef's structure.

Coral eat algae via their tentacle-fringed mouths.

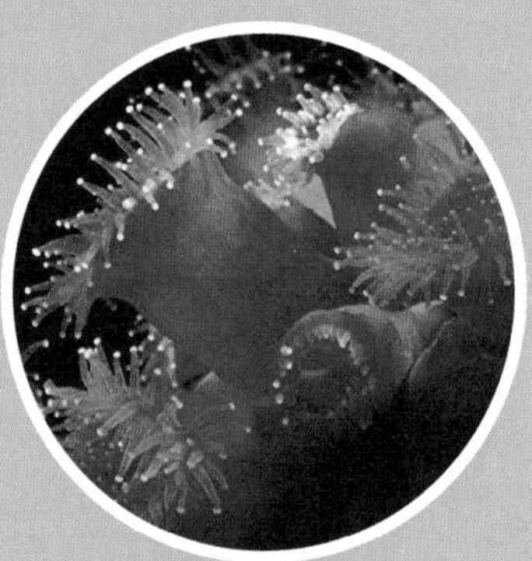

How atolls are formed

FRINGING REEF
Coral start to grow in the waters surrounding a volcanic island, forming a fringing reef.

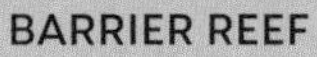

BARRIER REEF
The volcano starts to subside, creating a deepening pool of water around it. This creates a barrier reef.

 Island 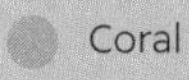 Lagoon  Coral Volcano

Different corals grow at different speeds, all of them relatively slow. Massive corals, for example, grow at a rate of 3–20 millimeters per year, while branching corals can grow up to 4 inches (100 mm) per year.

Charles Darwin proposed a theory to describe how rings of coral reefs, called "atolls," are formed across time. It all starts when free-swimming coral larvae attach themselves to underwater rocks around the edges of islands or continents. Then, as the coral grows, the reef takes on various forms, morphing from a fringing reef to a barrier reef, and eventually into a lagoon-filled atoll.

Staghorn corals, named for their antler-like appearance, are thought to have evolved around 60 million years ago.

100,000–300,000 years

DEVELOPING ATOLL

The volcano has almost disappeared. All the while, the coral is growing up toward the light.

ATOLL

The volcano is below sea level, leaving behind a ring of reef at the surface. With its circular internal lagoon, the atoll has formed.

CORAL BLEACHING

Coral reefs occupy less than 0.1 percent of the ocean floor yet house 25 percent of all known marine species. This makes them one of the most biodiverse ecosystems in the world. They provide homes, food, and spawning grounds for millions of species, including tropical fish, cuttlefish, and crustaceans. Around the world, they are a source of food and income for coastal communities and act as natural breakwaters, protecting thousands of miles of coastlines from storms and erosion.

As our species' influence on the planet intensifies, these vitally important ecosystems now find themselves in danger. Over the last 30 years, half of the Earth's tropical coral reefs have vanished because of pollution, overfishing, and unsustainable coastal development. Now climate change is adding to their woes.

To remain healthy, coral depend on tiny microscopic algae, called "zooxanthellae," which live inside their cells. It's a symbiotic relationship. The photosynthetic algae provide the coral with nutrients. When the temperature rises, the algae begin to produce toxic molecules, prompting the coral to eject them into the surrounding water. The algae give corals their colorful hues, so when they are gone, the corals appear ghostly, or "bleached." A tiny change in temperature, only 1.8°F (1°C), is enough to trigger bleaching. If the water then cools quickly, within a couple of weeks, the algae can recolonize the corals. But if the high temperatures persist or happen too frequently, the coral die.

Bleaching events are now happening with increasing frequency. As the coral reefs die, so, too, do the life-forms that they support. Even if climate change is limited to a temperature increase of 1.81°F (1.5°C), it's estimated that 70 percent of the world's corals will be lost by 2050. If this rises to 3.6°F (2°C), almost all the world's reefs will disappear in the same time span.

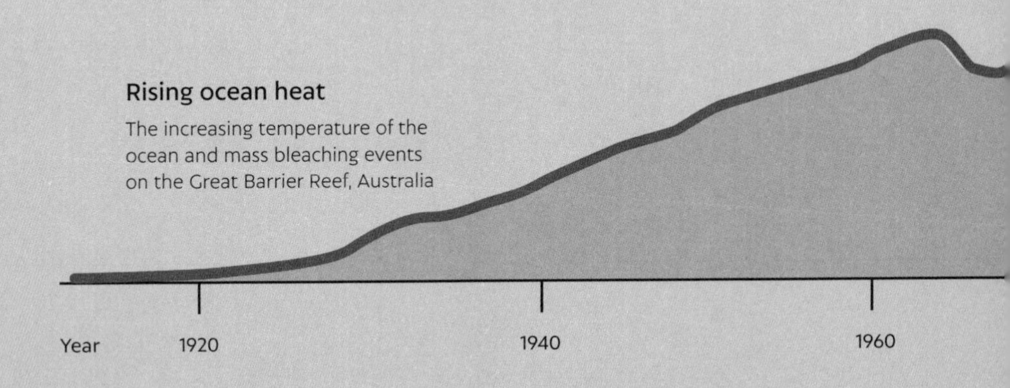

Rising ocean heat

The increasing temperature of the ocean and mass bleaching events on the Great Barrier Reef, Australia

GREAT BARRIER REEF

The Great Barrier Reef, located off the coast of Australia, is the world's largest coral reef system and is around 500,000 years old. It occupies an area of around 131,275 square miles (340,000 sq km). Like other reefs around the world, global warming has had devastating affects on this ecosystem. The reef is 1.81°F (1.5°C) warmer than it was 150 years ago. It has experienced 6 mass bleaching events in the last 24 years because of rising sea temperatures, and now these devastating episodes are occurring more frequently.

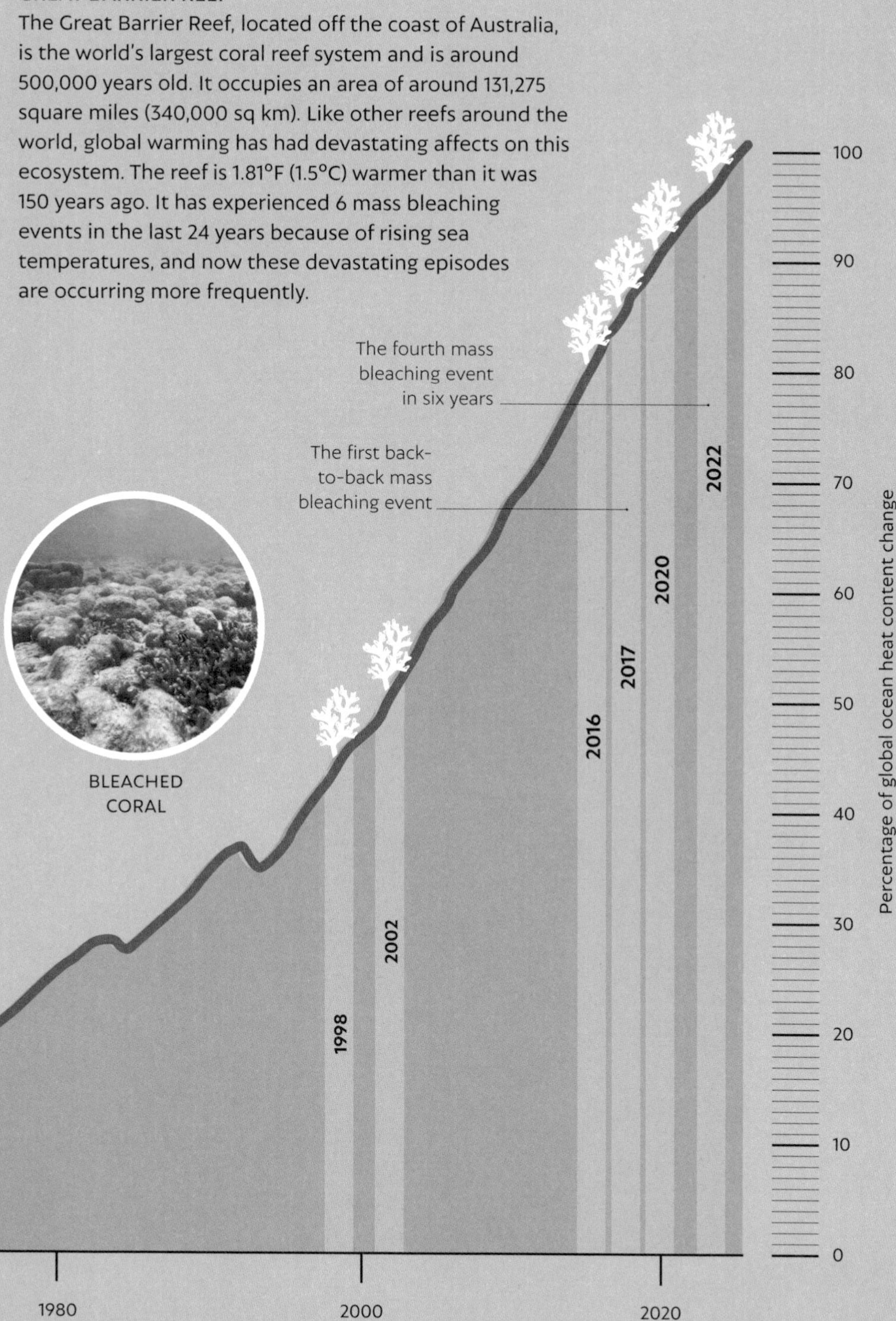

SPROUTING UP

Some plants seem to grow tall in no time at all, while others seem to never change. Plant growth is influenced by many factors, including location, climate, light, rainfall, and age. Younger plants, for example, tend to grow faster than older plants, and plants in warm, wet environments tend to grow faster than plants in cold, dry places.

Many succulents are extremely slow growing. This is an adaptation to life in environments where resources are scarce. The Saguaro cactus grows in the parched Sonoran Desert of Arizona, and for the first eight years of its life, it only grows around 0.05 millimeter per day. After that, growth rates vary. The greatest growth period occurs when the cactus grows branches, but by this time the cactus is typically 50–100 years old.

Dodders, meanwhile, may not grow big and tall, but their tendrils can still grow at a rate of around 150 millimeters per day. Dodders are a group of parasitic plants that suck water and

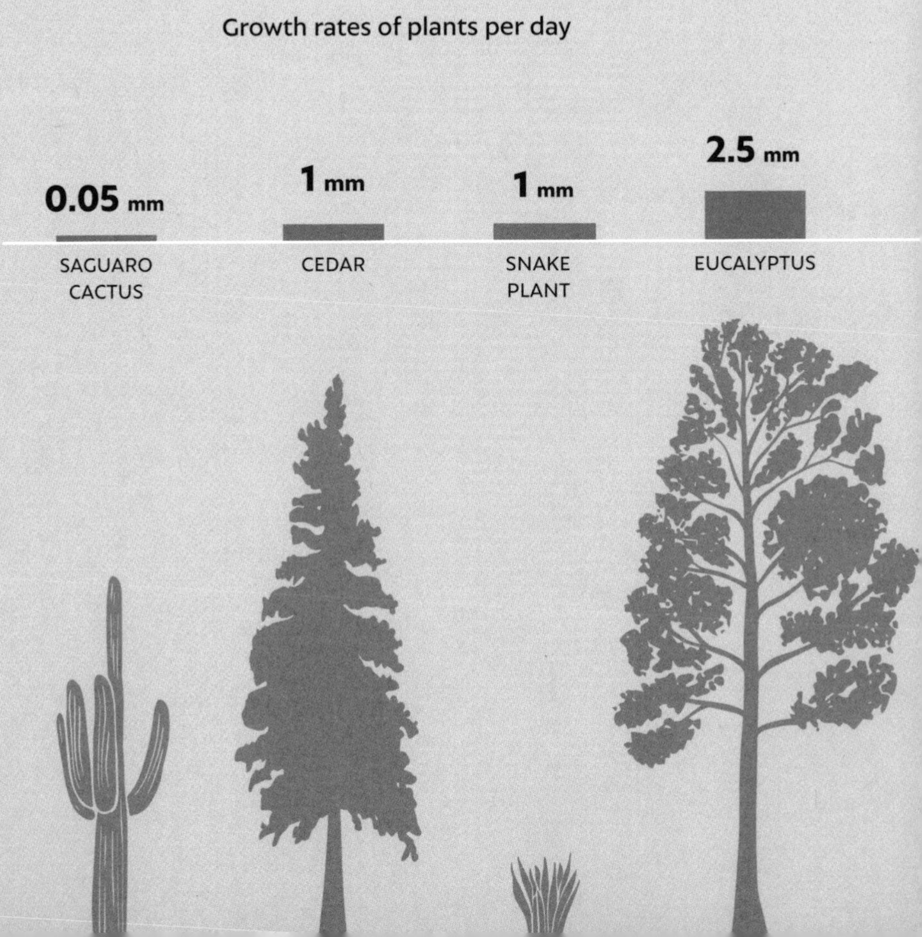

Growth rates of plants per day

0.05 mm
SAGUARO CACTUS

1 mm
CEDAR

1 mm
SNAKE PLANT

2.5 mm
EUCALYPTUS

nutrients from their hosts via tiny probing structures called "haustoria." They wrap themselves around their hosts and can smother entire trees. To achieve this rapid growth and minimize energy expenditure, they have no roots and their leaves are reduced to small, scalelike structures.

At the other end of the spectrum, notching up 1 meter (3 ft) of growth per day, Chinese moso bamboo is one of the fastest-growing plants on Earth. It grows in dense forests where there is intense competition for sunlight, so it has evolved to shoot up quickly. It achieves this by growing as a single stick, or shoot, connected underground to a parent plant by a stem called a "rhizome." The parent plant provides the shoot with nutrients, so it does not need to grow leaves of its own until it is fully grown. In addition, each shoot grows at a constant diameter, so unlike woody plants, it doesn't waste energy on producing growth rings.

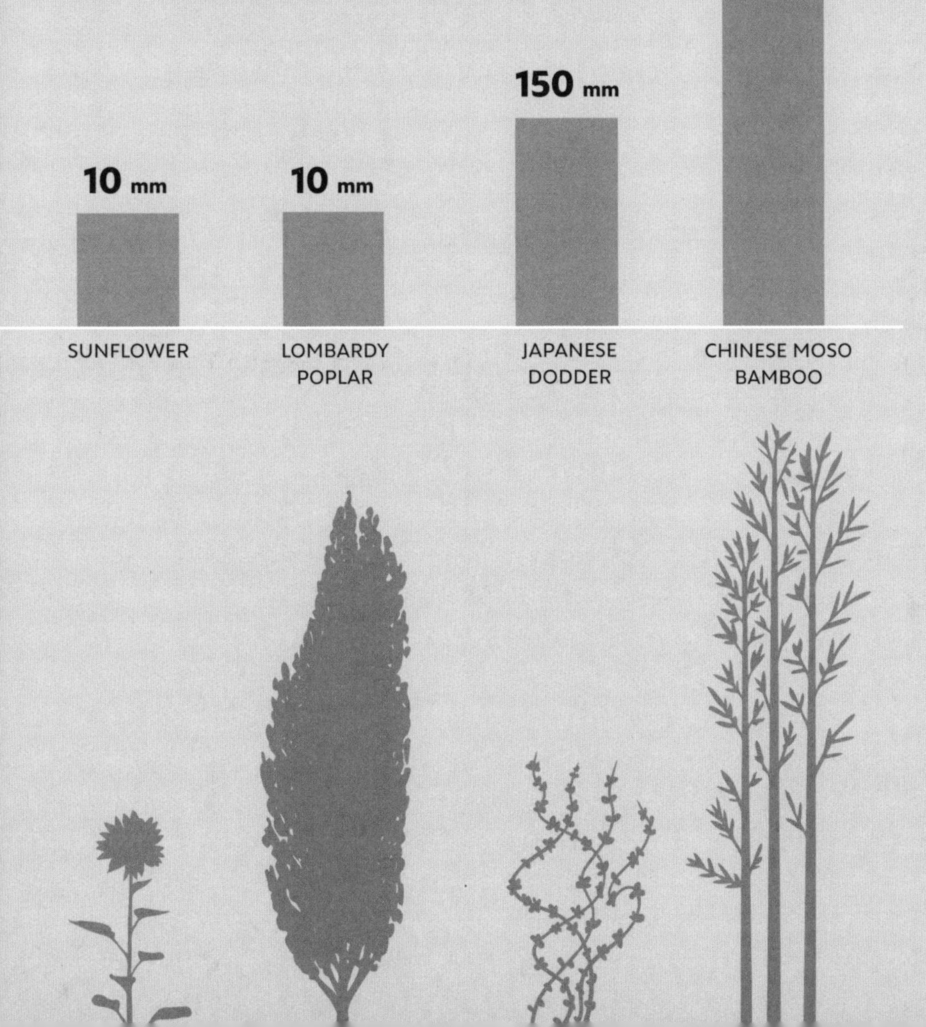

THE TURBO-CHARGED GROWTH OF POND WEED

All growth is relative. Compared with the daily growth of a big, tall bamboo plant, for example, the daily growth of a little bonsai tree may seem trivial. However, when size is taken into account and relative rather than absolute growth is studied, this chasm begins to shrink. Some of the most rapidly growing plants are actually very small indeed.

Anyone who has ever tried to keep a garden pond will be familiar with duckweed: the tiny, aquatic plant that quickly smothers the surface. Duckweed is among the most rapidly growing of all plants. A single floating frond of one species, *Wolffia australiana*, divides once every one or two days, enabling it to completely cover a pond within just a few weeks. Just like dodders, they achieve this turbocharged growth by skimping on roots and growing only the tiniest leaves. When researchers decoded its DNA in 2021, they found that *Wolffia* has fewer genes that are regulated by light and dark cycles compared to other plants. This means it can grow more at night, further accelerating its growth. It is a nuisance to pond keepers, but the plant has been shown to have a very high protein content, so it could prove useful as a future food source.

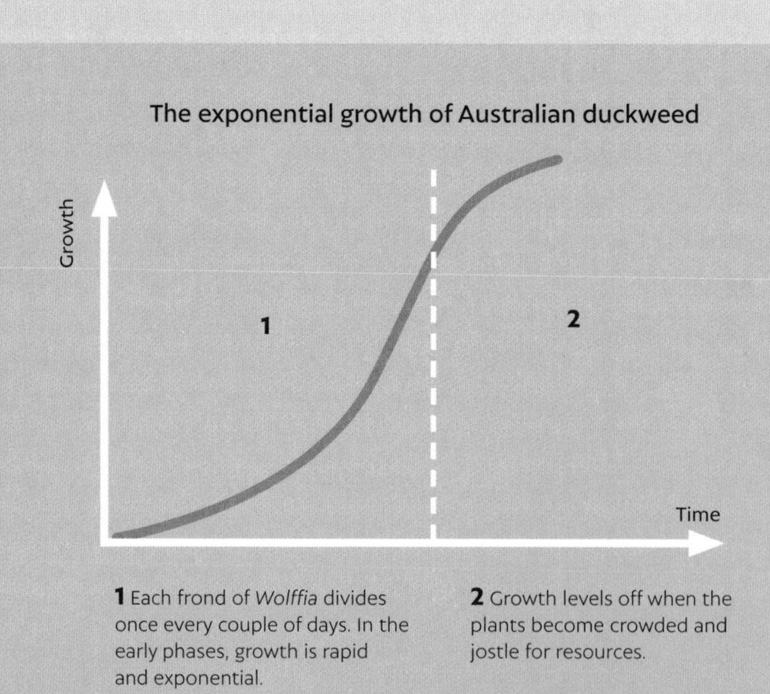

1 Each frond of *Wolffia* divides once every couple of days. In the early phases, growth is rapid and exponential.

2 Growth levels off when the plants become crowded and jostle for resources.

SPEEDING UP PLANT GROWTH

Scientists worldwide are trying to find ways of making crops grow faster. Agricultural fertilizers can make a big difference, but now researchers are trying to directly alter the process of photosynthesis.

During photosynthesis, plants use solar energy to turn carbon dioxide and sugars into valuable nutrients. Along the way, however, toxic compounds are produced, which the plant then has to recycle. This uses energy and limits growth. In 2019, scientists in the US used genetic modification to modify photosynthesis in tobacco plants. As a result, the plants produced fewer toxins and grew 40 percent larger than regular tobacco plants. Now researchers are working to boost the growth of other industrial crops, such as soybeans, rice, and potatoes.

GREEN TOBACCO PLANT

Duckweed is a common pond plant. There are different species, but the leaves of *Wolffia australiana* are less than 1 mm long.

REGENERATION IN ANIMALS

The human liver is unusual among our bodily organs. It can regrow to a normal size even after most of it has been removed. It's impressive but not invincible. Diseases such as cancer and hepatitis can damage it irreparably. Elsewhere in the animal kingdom, however, real-life "wolverines" do exist.

Top of the list is the axolotl: the amphibious salamander from Mexico. They can regrow parts of organs, such as the heart and spinal cord, as well as entire limbs. They retain this ability throughout their lives and the replacements are exact copies of the originals. Lizards, in contrast, are unable to regrow their arms and legs, but can regrow their tails. When this happens, the new limb serves the same function as the original, but it is a simplified version and is produced via a different mechanism.

Limb regeneration in the axolotl

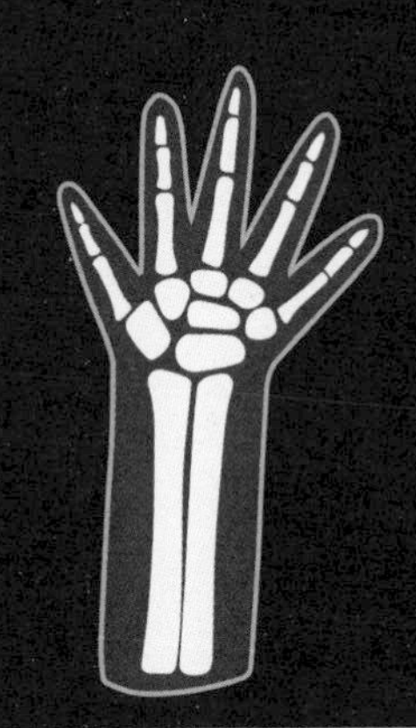

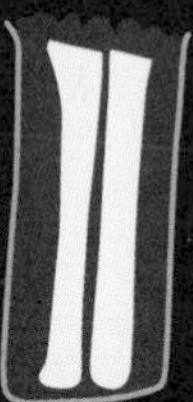

MINUTES

HOURS TO DAYS

FUNCTIONAL LIMB
Axolotls have four toes on the front feet and five toes on the back feet. They contain a variety of cells, including skin, bone, and cartilage.

TRAUMA
A limb is severed.

BLOOD CLOTTING
The wound is plugged by cells called "platelets" and strands of a protein called "fibrin." This causes bleeding to cease.

WOUND HEALING
Skin cells divide and cover the stump. This covering is called the "wound epidermis."

Limbs are made up of a variety of tissue types, including bone, cartilage, and muscle, which are organized in a very precise way. If a limb is to be regenerated, each of these tissues has to be produced and then fall neatly into place. Limb regeneration in the axolotl progresses through a series of distinct stages.

- Skin
- Bone
- Cartilage
- Muscle
- Stem cells

REGENERATIVE THERAPIES

We can learn much from the axolotl. By studying the cellular and molecular changes that accompany its limb regeneration, scientists hope to develop regenerative therapies for people. They have realized, for example, that a type of immune cell called a "macrophage," plays a key role. If macrophages are eliminated, then the severed limb just scars over and does not regrow, so therapies that boost macrophage activity may prove useful.

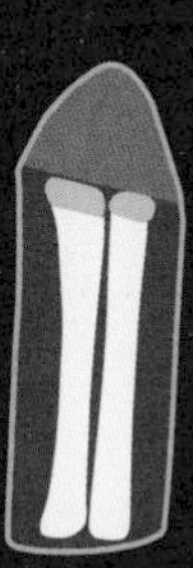

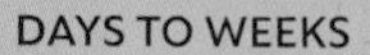

WEEKS

1 MONTH AFTER TRAUMA

BLASTEMA FORMATION
A cone-shaped structure called a "blastema" forms. Inside this, mature cells such as bone, cartilage, and muscle revert to a more youthful state, in which they can divide and generate new tissue.

LIMB REGROWTH
New skin, bone, cartilage, and muscle are generated. Digits start to form, and the limb regains its original shape.

FUNCTIONAL LIMB
The growing structure develops into a perfect copy of the lost limb. It is connected to the rest of the body via nerves and blood vessels.

LIMB REGENERATION IN INVERTEBRATES

Many invertebrates (animals that lack back bones) also have impressive regenerative abilities. Spiders, starfish, sea cucumbers, and flatworms can regrow parts and sometimes all of their body.

FLATWORM

TIME SPAN: 4 weeks

Flatworms are a group of tiny, flat, soft-bodied worms. When diced into pieces, each piece can generate a new worm in just a few weeks. They have a strong sense of head and tail, or "polarity." The "head end" of a segment will always sprout a new head, while the "tail end" will always sprout a new tail.

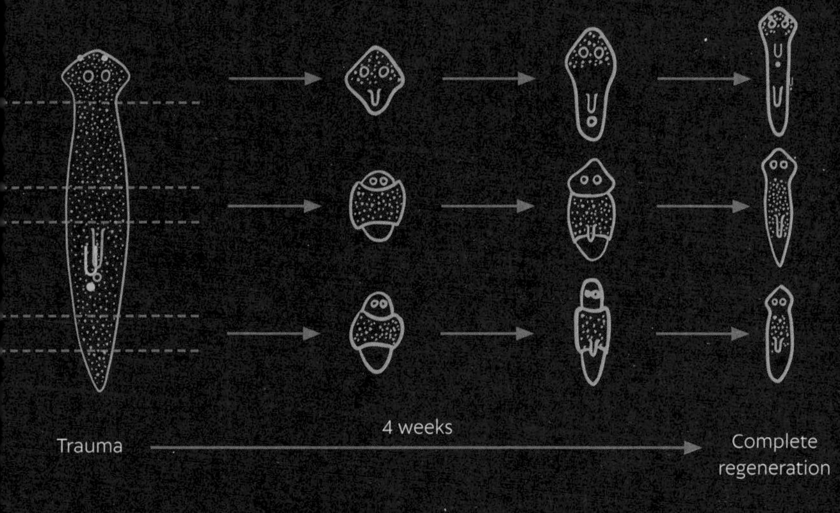

SPIDERS

TIME SPAN: Days to months

Spiders have hard exoskeletons, so they must molt in order to grow. They go through a finite number of molts in their lifetime. If a leg is severed while one or more molts remain, then the limb can regrow. Often, the new leg is thinner and shorter than the original, and it can take two or three molts before the regenerated limb matches the original.

STARFISH

TIME SPAN: Up to a year

Starfish can regrow arms and sometimes entire bodies. There are around 2,000 species of starfish. They can achieve different degrees of regeneration in different ways.

The severed starfish retains the central disk and a large part of its body. It can eat and move as it regrows its two missing arms.

A Each arm contains a copy of vital organs, some light-sensitive cells, and tiny projections called "tube feet," which enable the animal to move.

B The central disk contains the mouth and digestive system.

A Arm

B Central disk

A single severed limb has retained the central disk with its mouth and stomach. It can still feed and digest food as it regenerates the rest of its body.

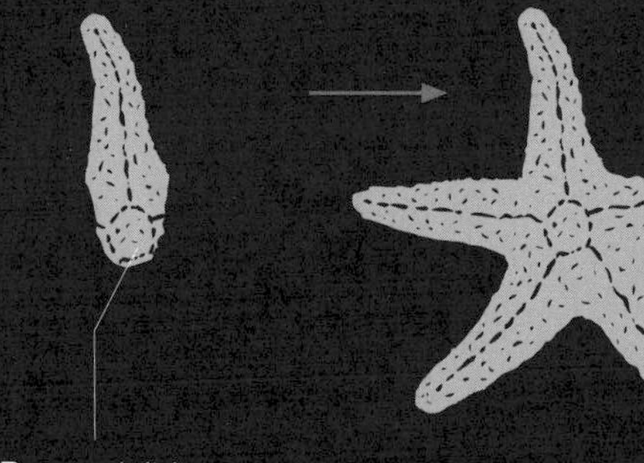

A single severed limb is left, without its central disk and mouth. In a few tropical species, regeneration can still occur. The animal survives by using nutrients that are stored inside the limb, until the central disk regrows.

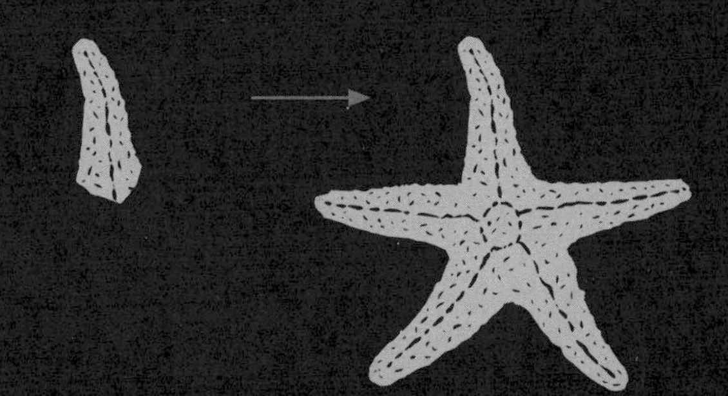

5
BEHAVIORAL SPANS

INTRODUCTION

If there's one lesson to learn from this book, it's this: Never stand behind a hippo. Hippos have an eccentric habit of urinating and defecating at the same time, while spinning their tail like a fan to propel their fragrant gifts as far and wide as possible. Males do this to assert their dominance, mark territories, and win a mate—and they say romance is dead! Dung showering, as it's called, is particularly pronounced during the mating season, which usually occurs between February and August, when the females reach peak fertility for a short three-day window.

This is one example of the rich and varied behavior that is commonplace in the natural world. In the sky, insects and birds use powerful air currents to assist with epic migrations that can take months. In the water, male water bugs carry their unhatched eggs on their backs for weeks to protect them, while a male pufferfish may spend days creating a circular, sandy artwork to attract a mate. And let's not forget that plants exhibit behavior, too. As they grow toward the light, plants release chemicals to communicate with their neighbors, and they snap together specialized leaves to catch insects. Plants react to their changing environments.

Behavior is the way that living things respond to their environment, so all living things "behave." Some of this behavior is preprogrammed at a genetic level; it is inherited and known as "innate" behavior. Spiders, for example, instinctively know how to make webs. They don't need to be taught.

Meanwhile, "learned" behaviors are not present at birth and have to be taught. Meerkats, for example, spend hours teaching their pups how to handle dangerous scorpions. Let's discover the wonderful world of animal and plant behavior.

Meerkat pups learn how to forage from their parents, who sometimes present them with dead scorpions.

SLEEPING HABITS

From fruit flies to humans, most creatures need rest, but the length and type varies hugely between species. Out of all mammals, koalas need the most sleep (up to 22 hours a day), while giraffes sleep the least (fewer than 30 minutes per "nap"). Pythons sleep for 18 hours, but scientists wonder if bullfrogs rest at all. Swifts rest on the wing, allowing half their brain to sleep, while the other half remains alert. Sleep restores glycogen, a form of glucose that aids brain function. Research shows that humans, rats, flies, and cockroaches all die if sleep-deprived. The record for human sleeplessness is 264 hours, achieved in 1964 by an American named Randy Gardner.

CATS 16–18 hours

SPERM WHALES 10–15 minutes
Sperm whale sleep bobbing at the surface of the water in an upright position. Scientists suspect that they require the least sleep among mammals, accounting for less than 10 percent of their daily activities.

KOALAS 22 hours
These marsupials are one of the sleepiest animals on Earth.

ARMADILLOS 16 hours

FRIGATE BIRDS 45 MINUTES
When flying, frigate birds sleep up to 45 minutes a day in 10 second bursts. On land, they sleep a minute at a time for up to 12 of every 24 hours.

RATS 13 hours

Hours

9:00 10:00 11:00 12:00 PM 1:00 2:00 3:00 4:00 5:00 6:00 7:00 8:00

DOLPHINS 8 hours

Dolphins sleep on the water's surface and use only half their brain to rest, which is called "unihemispheric sleep." Half their brain must stay awake to be alert to danger and in order to breathe, as breathing is not automatic.

GIRAFFES 4–5 hours

In captivity, giraffes sleep mostly at night. In the wild, they sleep briefly but intensely, either standing up or lying down.

ANIMAL MIGRATIONS

Sometimes the smallest animals can make the greatest journeys. Many animals migrate across vast distances and over long periods of time in order to find vital resources such as food and mates. The painted lady butterfly is one of them. It weighs less than one gram, has a brain the size of a pinhead, has no opportunity to learn from older generations, and yet manages to achieve a remarkable feat.

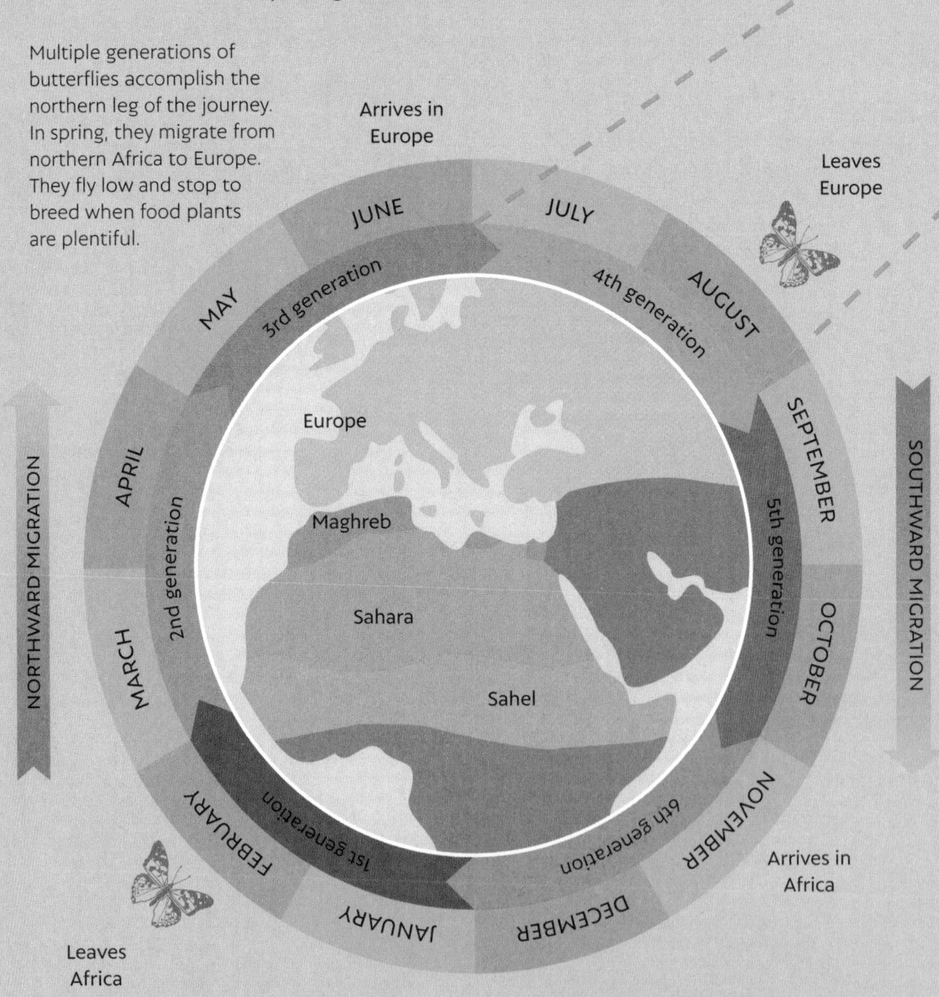

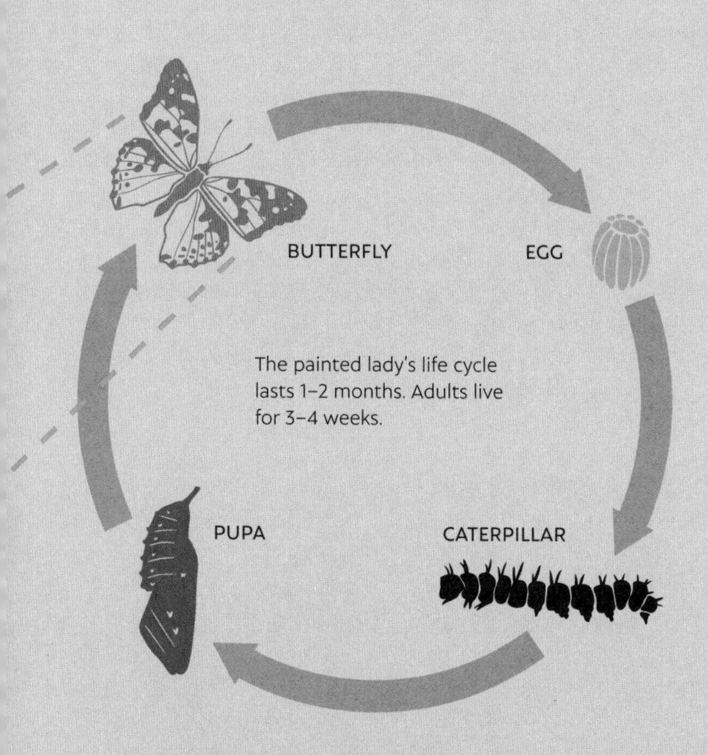

PAINTED LADY

Every year, the painted lady butterfly makes a 9,010-mile (14,500 km) round trip from tropical Africa to Europe and back. It is a multigenerational migration. This means that the complete trip is not made by any individual butterflies but by up to six successive generations, one after the next. The migration is performed so that the butterflies can find food for their offspring to eat.

Scientists have known about the painted lady's northward migration for decades. The butterflies frequently come to rest and are easily observed during this part of the journey, but no one knew what happened to the butterflies after they arrived in Europe. Some thought they simply died there. Studies have now shown this is not the case. For a long time, the southward migration went unnoticed because it occurs at high altitudes and the butterflies cannot be seen from the ground. They fly at heights of over 1,640 feet (500 m) and can reach speeds of around 31 miles (50 km) per hour by making the most of suitable wind currents.

MORE AMAZING MIGRATIONS

Insects are not the only animals to migrate. Many birds, mammals, fish, reptiles, and crustaceans also partake in migration. These epic journeys take place in the air, on the land, and in water, and they all occur across different durations and locations.

Duration of migration

10 months

COMMON SWIFT

With the exception of two months in Europe that are set aside for hatching and raising their chicks, life for the swift is one long migration. They are on the wing, almost constantly, for the 10 months it takes them to leave Europe, fly to Africa, and then return to Europe again. They feed, mate, molt, and sleep in the sky. Swifts can live up to 21 years, so a single individual may fly more than 620,000 miles (1,000,000 km) in its lifetime.

2–4 years

SALMON

Salmon migrate in the opposite direction to eels. Salmon spend most of their adult lives in the ocean, but they return to fresh water to spawn. Journey lengths vary, but some salmon take several years to travel the many thousands of miles back to their spawning ground.

2–12 months

EEL

Eels migrate in the opposite direction to salmon. They travel from their adult homes in European freshwater rivers, downstream to their spawning grounds in the Sargasso Sea. It's a journey of more than 2,980 miles (4,800 km). Tagging studies reveal that while some eels take the quickest and most direct route, others take a longer, more meandering journey and spawn in subsequent seasons.

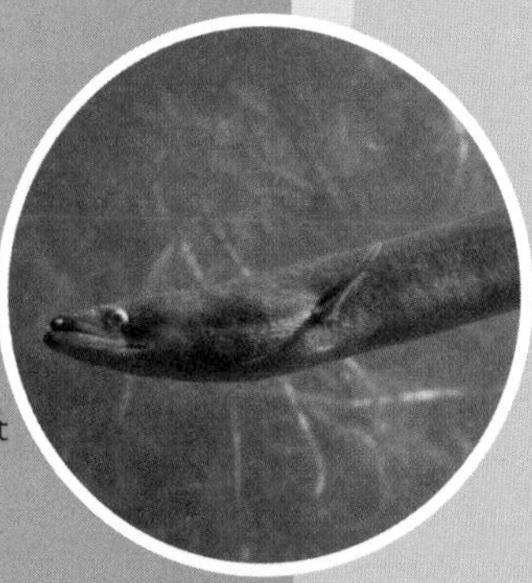

1–2 months

HUMPBACK WHALES

Humpback whales live in oceans all over the world. They have many migratory routes, always moving between cooler, krill-rich feeding grounds and warmer breeding grounds. In the north Pacific, some humpbacks migrate 3,100 miles (5,000 km) from Alaska to Hawaii in as little as four weeks. Meanwhile, at the opposite pole, humpbacks take a couple of months to migrate from their feeding grounds in Antarctica to their breeding grounds along the southern, eastern, and western coasts of Australia.

2 days–2 weeks

SPINY LOBSTER

With the shallow California waters cooling around them, spiny lobsters line up, conga-style, and march single file toward the warmer waters of the deeper ocean. Lobster chains can contain dozens of animals, all linked up, antennae to tail.

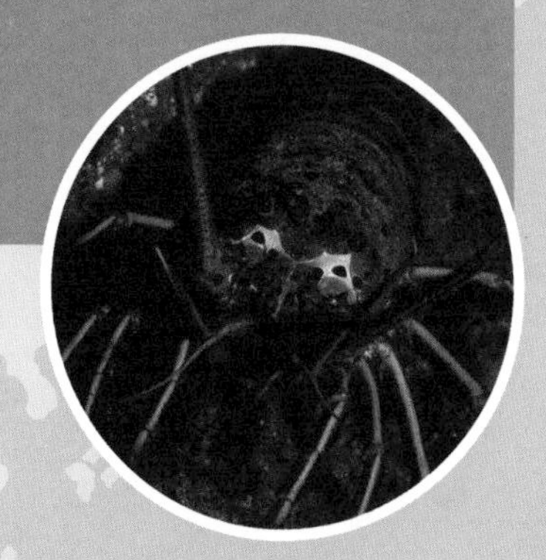

2 weeks

DUSKY GROUSE

Dusky grouse live in the highlands of North America. In spring, they take a couple of weeks to shuffle and flap a few hundred feet down the mountain, from coniferous to deciduous woodlands, where they go on to breed. It's one of nature's shortest migrations, both in terms of time and distance.

NO TIME TO DIE

The assassins of the natural world are both devious and deadly. Parasites and predators have evolved many different styles of killing. Some endings are slow and lingering, while others are mercifully brief. If the thought of a pet cat beheading a mouse makes your stomach churn, then just wait until you hear the one about the killer fungus that turns its prey into a zombie.

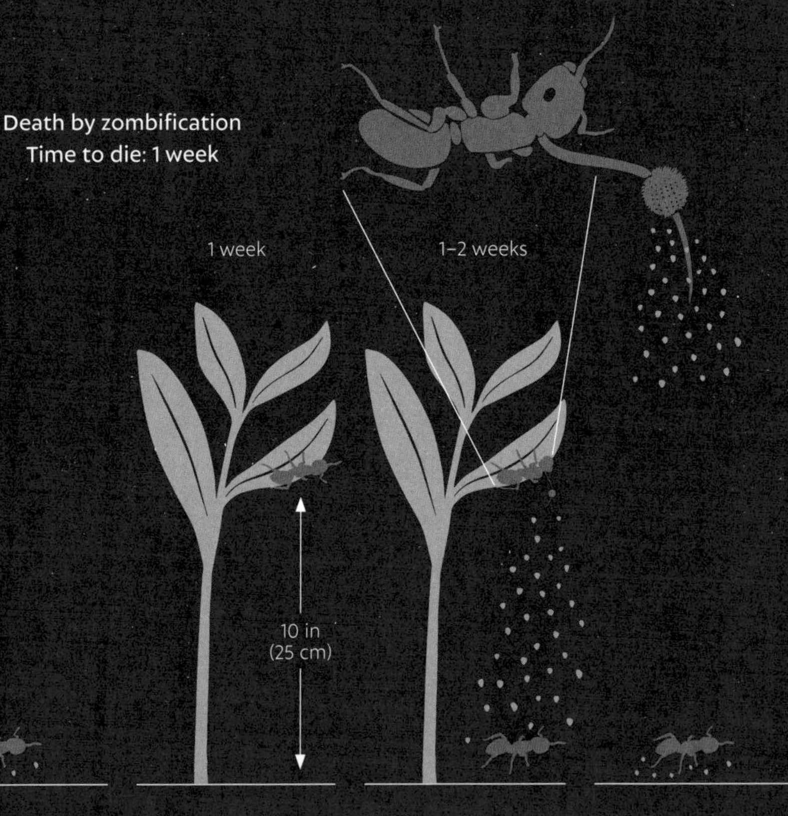

INFECTION
A forager carpenter ant walks through a patch of forest floor that is infested with fungal spores. The spore releases an enzyme that eats through the ant's outer shell. The fungus enters the ant.

MIND CONTROL
The fungus directs the ant's behavior. It makes the ant climb a nearby plant and bite into a leaf that is 10 inches (25 cm) above the ground, where it is the perfect temperature and humidity for the fungus to grow. The ant then dies.

EVISCERATION AND SPORE PRODUCTION
The fungus eats the ant's internal organs. It extends a long stalk out of the ant's head, which then produces spores that fall to the forest floor.

INFECTION
The spores lie in wait for the next passing ant. When the ant walks over the fungus, the cycle begins again.

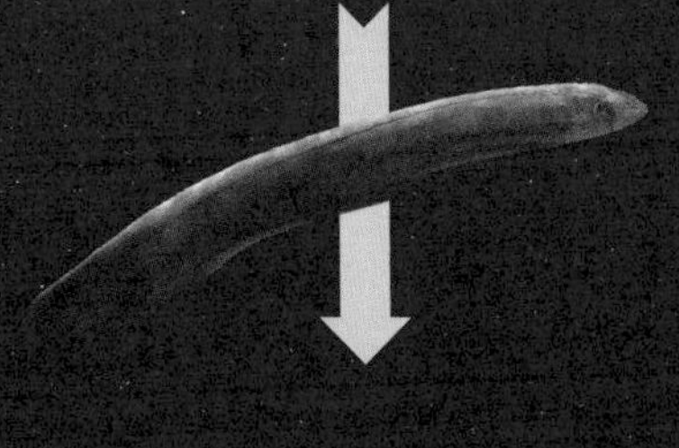

DEATH BY ELECTROCUTION

Time to die: Milliseconds

The tail of an electric eel can produce a 600-volt jolt of electricity that stuns and immobilizes its prey. It also sends out intermittent jolts that cause the muscles of potential victims to twitch involuntarily, thereby revealing their location.

DEATH BY TONGUING

Time to die: Less than a second

Chameleons lash out their long, super-sticky tongues to capture insects in less than a second. A chameleon's tongue can go from 0 to 59 miles (95 km) per hour in one-hundredth of a second, which is faster than a sports car.

DEATH BY STABBING

Time to die: Seconds

Loggerhead shrikes, also known as "butcher birds," have sharply hooked beaks that they stab into the head or neck of their prey. They then shake their victims vigorously, causing whiplash-like injuries, and they store any uneaten prey by impaling it on a thorn or bit of barbed wire.

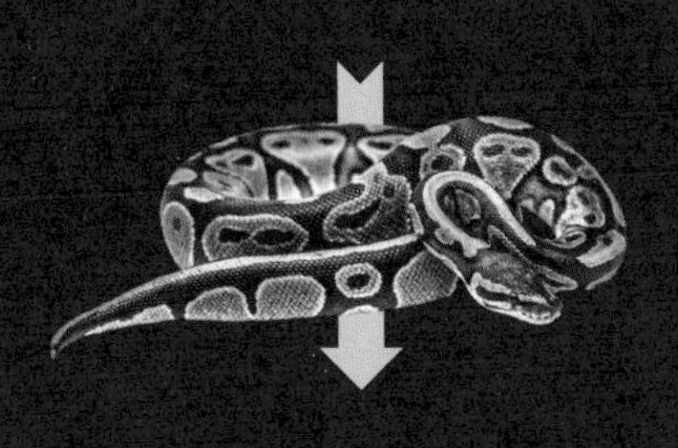

DEATH BY SQUEEZING

Time to die: Minutes

The boa constrictor squeezes prey so hard that it can no longer pump blood around its body. Consciousness is lost within seconds, and death follows minutes later, as the heart and other vital organs shut down.

DEATH BY VENOM

Time to die: Minutes to hours

Snake venoms can kill in different ways, such as paralyzing the nervous system or causing hemorrhaging. The black mamba has the fastest-acting venom of any snake. It can kill a human in 20 minutes. The venom of the western diamond rattlesnake kills more slowly. Its venom contains proteins that enable the snake to track down bitten prey.

KILLER PLANTS

Animals and fungi aren't the only deadly species out there. More than 500 plant species attract, trap, and kill prey such as insects. They grow in nutrient-poor soils or putrid waters, where the strategy has evolved as a way to supplement their diet. Carnivorous plants are thought to have evolved independently at least 12 times, and there are a handful of different trap styles. The speed of each trap varies, and the plant then digests its prey over hours and days.

BLADDERWORT PLANT

BLADDER TRAP

Time to ambush: Half a millisecond

The underwater leaves of bog-dwelling bladderworts are bladder-shaped traps. Each trap has a tiny door surrounded by fine, touch-sensitive hairs.

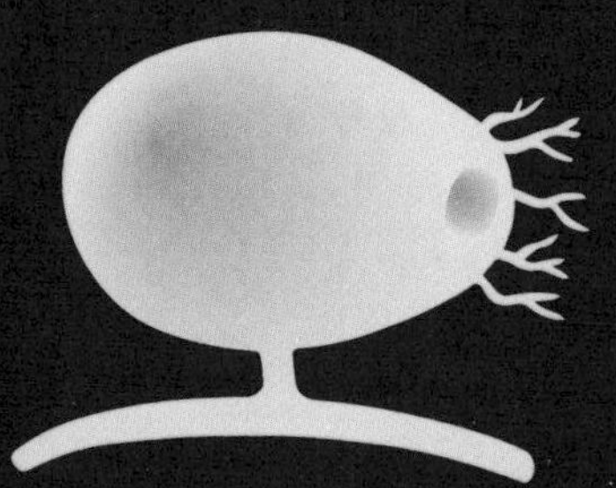

1 hour

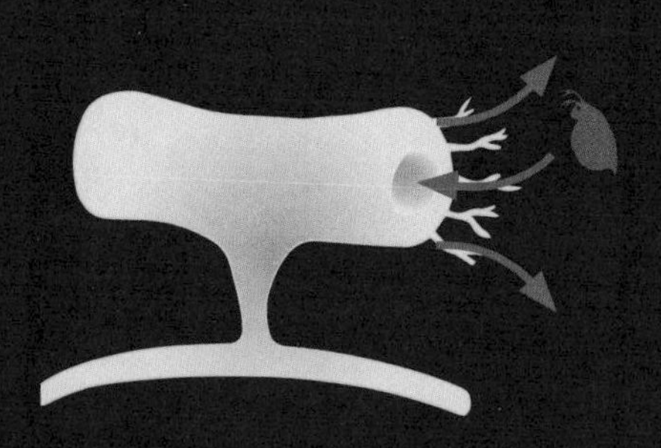

Glands in the trap pump out water. Air inside the trap is at a lower pressure than the surrounding water, and the door bulges out. A tiny crustacean swims too close.

0.5 millisecond

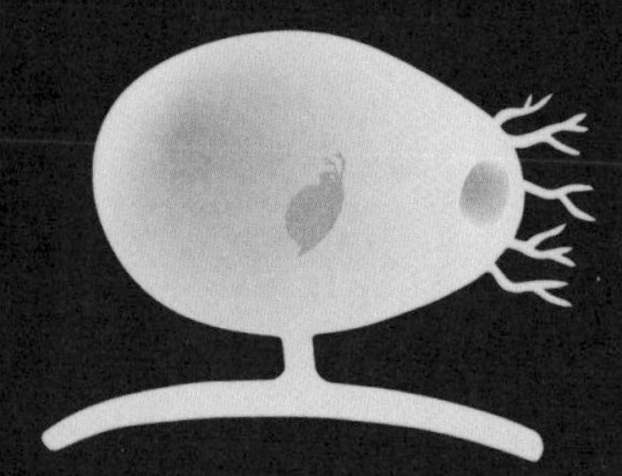

The prey touches the tiny hairs on the outside of the door. The door opens and water and prey are sucked in with accelerations of up to 600G (G-force). The door shuts. The crustacean is then slowly digested over the next few hours.

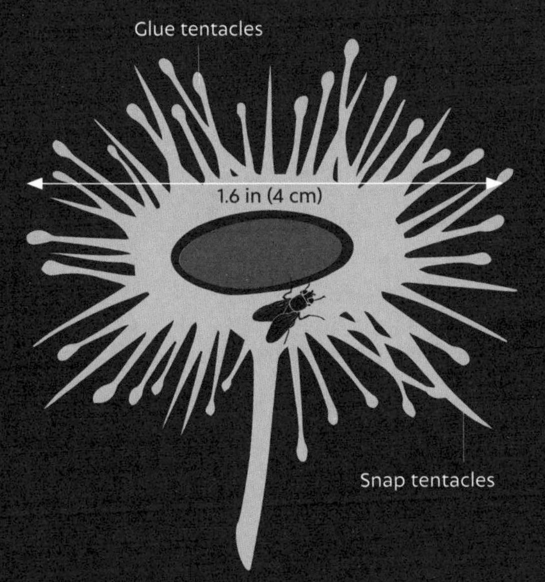

COMBINATION TRAP

Time to ambush: 75 milliseconds

The Australian sundew plant combines a sophisticated catapult trap with a glue trap. German researchers, who filmed the plant in extreme slow motion, have shown how the prey is ensnared. Insects walking near the edge of the sundew plant trigger long, touch-sensitive snap tentacles that catapult their prey onto nearby sticky glue tentacles in less than a millisecond. They are then drawn inside the trap and digested by enzymes.

AUSTRALIAN SUNDEW PLANT

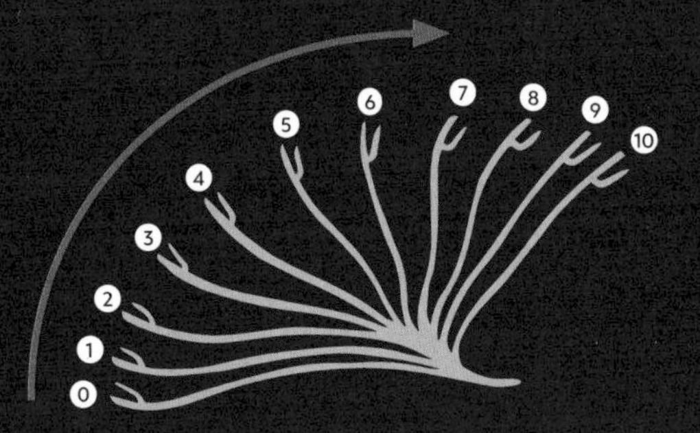

Snap tentacle in slow motion: Whiplike tentacles reach speeds of up to 56 feet (17 m) per second.

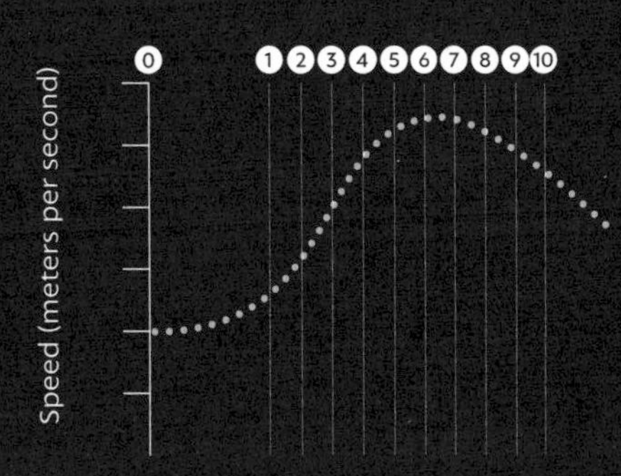

ALBANY PITCHER PLANT

PITFALL TRAP

Time to ambush: Minutes to days

As well as its regular-shaped leaves, the carnivorous Albany pitcher plant from Western Australia has modified leaves that are used to catch prey. These are moccasin-shaped and full of digestive juices. Insects are attracted to the plant by the nectar it produces.

A lid prevents rainwater from entering the modified leaf, so the digestive juices are not diluted.

Spikes at the entrance allow the insect to enter but then hinder its escape.

THINK FAST

A mouse escapes from a predatory bird. A mountain hare runs down a steep precipice. Often, in the natural world, survival comes down to the ability to think fast. Delayed reaction times cost lives and yet studies show that even the fastest animal reflexes are surprisingly slow.

In a study from 2018, scientists from Canada's Simon Fraser University measured the reflex speeds in mammals of different sizes, ranging from tiny shrews to enormous elephants. They found that larger animals have slower reflexes than smaller ones.

This is not surprising. Although nerve impulses travel at about the same speed in different-size animals, they have farther to travel in bigger animals, so it takes longer. Reflex times vary from 10 milliseconds in small animals, such as mice, to 100 milliseconds in larger animals, such as giraffes. To put this in context, it takes less time for an orbiting satellite to send a signal to Earth than it does for a giraffe's spinal cord to send a signal to its foot.

It's a serious flaw, so larger animals have evolved various strategies to help compensate, such as moving slowly to allow them more time to respond to disturbances. Researchers also believe larger animals rely on anticipating future movements, which enables the animals to adjust their actions accordingly.

Reaction times in land-living mammals

Large animals have slower reflexes than small ones.

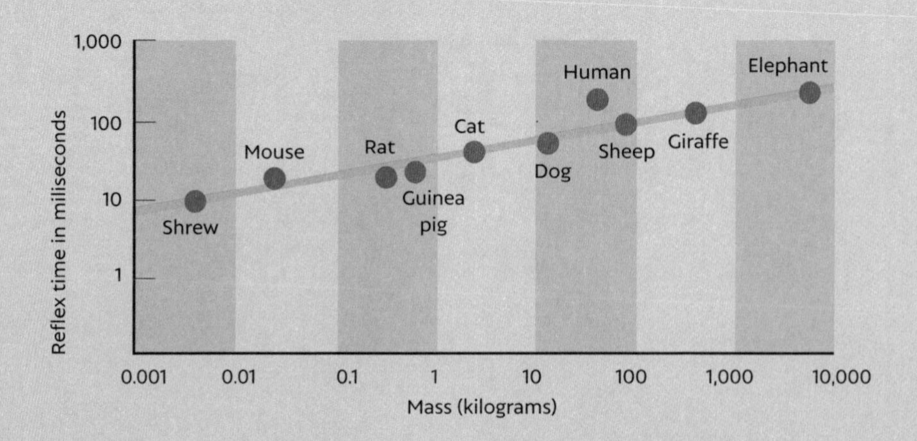

The mechanics of reaction

When a vertebrate like a giraffe reacts to a stimulus, such as a painful thorn underfoot, reflexes kick in. Without thinking, the signal is quickly detected and processed. A reaction is generated and the giraffe lifts its foot.

1 A giraffe treads on a painful thorn. The signal is sensed by receptors in the foot.

2 It travels along a sensory neuron, up the leg to the spinal cord.

3 It "jumps" from the sensory neuron to the relay neuron.

4 The signal passes along the relay neuron.

5 It "jumps" from the relay neuron to the motor neuron.

6 It travels along the motor neuron, down the leg to a muscle.

7 The signal "jumps" from the motor neuron to the muscle.

8 The muscle reacts. It contracts and the foot is lifted up.

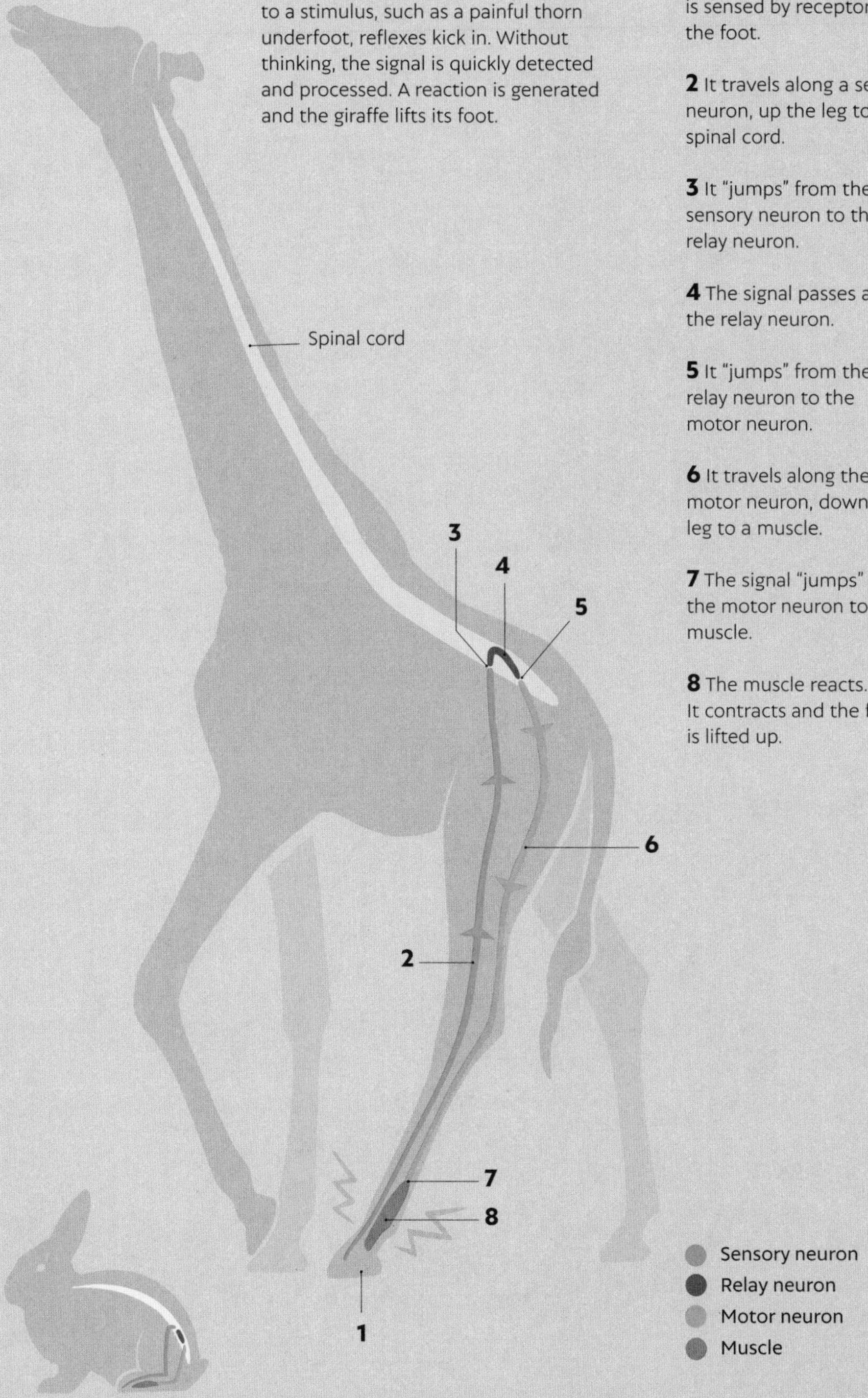

PARENTING STRATEGIES

From complete devotion to total neglect, there are many different parenting strategies in the natural world. The cuckoo, for example, is well known for abandoning its offspring to the care of others, while orangutans dedicate up to seven years looking after their young. But did you know that some insects also have parenting skills?

Around 1 percent of insect species show parental care. Male giant water bugs, for example, carry their unhatched eggs on

Parental care in the burying beetle

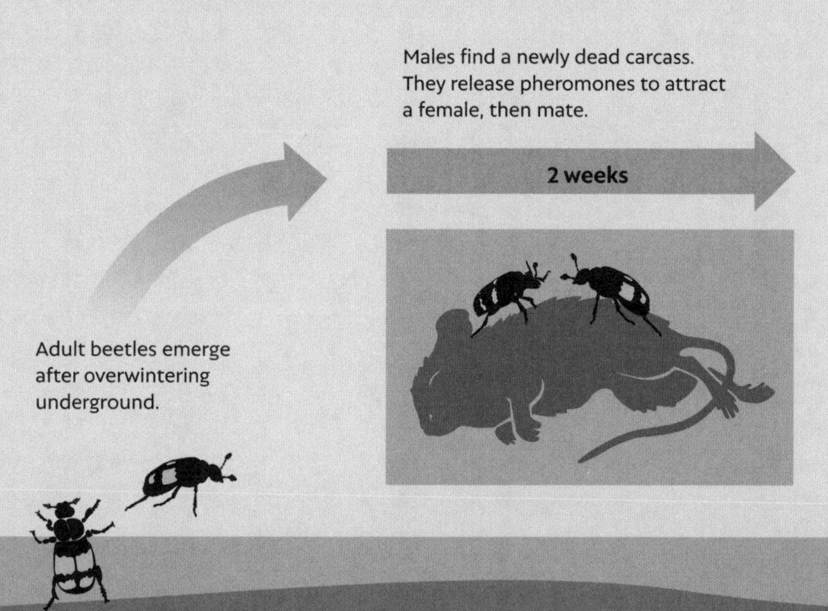

6 months

3 weeks

Adult larvae emerge. Many stay dormant in the soil over the winter.

The larvae pupate in the soil.

their backs for several weeks, until the larvae emerge. Female earwigs, meanwhile, fuss over their eggs for the week that they take to hatch. They groom and coat them in bacteria that have antimicrobial properties.

Burying beetles take things a step further. Both parents are involved in raising the offspring. They construct an elaborate underground larder and then feed their begging larvae. It is hard work and takes more than a week.

Parental care begins even before the eggs are laid. The carcass is their offspring's future food source. The beetles prepare it by snipping the animal's fur and burying it.

The beetles roll the carcass into a ball, then slather it in decay-slowing secretions. The female lays her eggs.

2–3 days

The larvae emerge and start to beg. The parents feed them with regurgitated carrion.

14 days

The larvae disperse into the surrounding soil.

PARENTAL CARE IN PLANTS

It is hard to imagine that plants look after their young, but some species do. Take the desert cactus (*Mammillaria hernandez*), a plucky little plant from Mexico. Adult cacti release an initial batch of seeds but keep part of the batch back, which is then stored for a year inside the "mama" plant's stem.

To count as parental care, there has to be some benefit for the retained seeds. Scientists tested this by comparing retained seeds against newly produced seeds that had not been retained. Seeds that had been "looked after" by a parent for a year were more likely to germinate and survive than young seeds that were released without this care. Nurtured seeds were also less likely to be eaten by predators or rotted by microorganisms. It might not have a nervous system or a "will" to protect its offspring, but the desert cactus really does seem to be looking after its future seedlings.

The practice of retaining mature seeds for extended periods of time is called "serotiny." It's common in desert-dwelling plants, for which conditions can be unpredictable, and it's seen as a bet-hedging strategy; if the parent plant does not produce offspring in the first year, it can always try again in the second. Many conifers also practice serotiny.

VIVIPARY

Other plants have evolved different ways to care for their offspring. Just like the desert cactus, red mangrove trees also retain their seeds, but in this case, they germinate and grow on the parent plant. The seedlings, "propagules," acquire nutrients from the parent, then grow down to the water and drop off the tree. This strategy is called "vivipary."

Seed survival in the desert cactus

Seeds that are looked after by their parents do better than seeds lacking parental care.

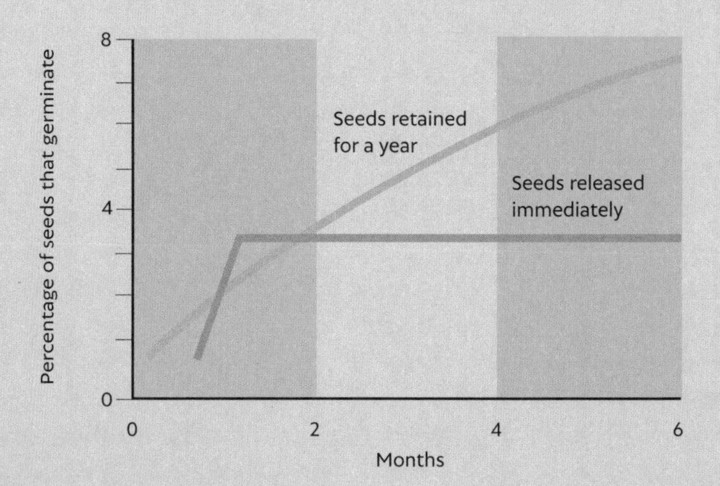

The desert cactus

In the desert, cactus seeds are nurtured inside the parent plant's stem.

HOME SWEET HOME

Our homes tend to be built from bricks and mortar, but for the rest of the animal kingdom, homes are built from a more eclectic range of materials, including mud, sticks, saliva, and air. These dwellings vary from simple depressions in the ground to sophisticated feats of engineering, and as they become more complicated, so, too, does the amount of time invested in them. While some are designed to be temporary, like the overnight slumbering pits of gorillas, others are designed for longevity. Termite mounds, for example, can last for thousands of years.

GORILLA
5 minutes
Gorillas use leaves, bushes, and branches to make 3-foot-wide (1 m) circular nests on the ground. Although they travel less than a mile in 24 hours, gorillas make new nests every day. Most gorillas sleep alone, although moms sleep with their young infants.

SIAMESE FIGHTING FISH
2–3 hours
Males make blobby nests from bubbles. They gulp down air, then release it in tiny, mucus-encased pockets. It takes thousands of bubbles to make a nest, which floats at the water's surface. Fertilized eggs are placed inside the nest, where they subsequently hatch.

PLATYPUS
8 hours
Platypuses live in burrows. After mating, a female will spruce up the burrow and build a nest at one end. She softens the floor with folded, wet leaves and prepares the nest from reeds and leaves. It's constructed in bursts of activity, spanning eight hours over three nights.

Nest building times in animals

Build difficulty

FLAMINGO
2–3 days
Flamingo pairs build their nests on mudflats. They use their beaks to mush up mud, leaves, and grasses, which they then use to make a little, bowl-shaped mound. Parents take turns to incubate the single egg, which takes about 30 days to hatch.

EDIBLE-NEST SWIFTLET
2 months
The translucent, cup-shaped nest of this little southeast Asian bird is made exclusively of solidified saliva. The substance is produced from a gland under the tongue, and it hardens on contact with air. Unfortunately for the swiftlet, its nests are the main ingredient in bird's nest soup.

BALD EAGLE
1–3 months
It's the scale that makes this nest so impressive. At 6.5–10 feet (2–3 m) wide and up to 13 feet (4 m) deep, the bald eagle's nest is the biggest in North America. The stick-built construction is reused by the pair, who top it up with around 20 inches (50 cm) of new material every year.

BEAUTIFUL BUILDS

Termites and tits are among the most skilled architects in the animal kingdom. They both solve the problem of making a safe place to raise a family in two completely different ways.

Penduline tits, found in Eurasia, Africa, and North America, spend up to three weeks each year, creating an exquisite nest in which to raise their young. It's a multistage process that sees an increasingly complex construction dangling from a small branch. The nest is woven from soft plant material and animal fibers, with only a hint of spiderweb for added stretch. It is about 10 inches (25 cm) long and is used for a single breeding season before being discarded.

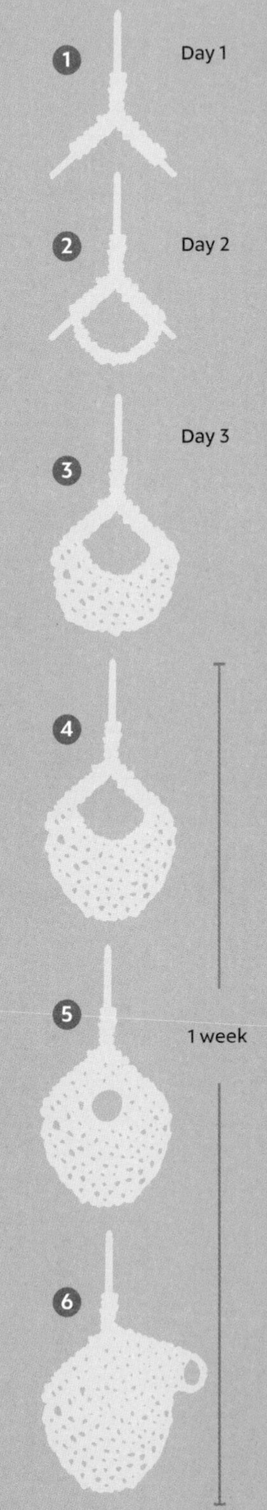

Penduline tit nest construction

1. The male wraps building material around a forked twig.
2. A rudimentary basket is created.
3. The male pauses construction for up to two weeks to find a mate. They continue the build together.
4. The basket is deepened and strengthened.
5. The basket is enlarged into a sphere with a hole.
6. The hole is enlarged to form a spout.

Some penduline tits also incorporate a false entrance leading to a false chamber. The entrance to the true nesting chamber is protected by hidden flaps, which are sealed shut with sticky spider silk.

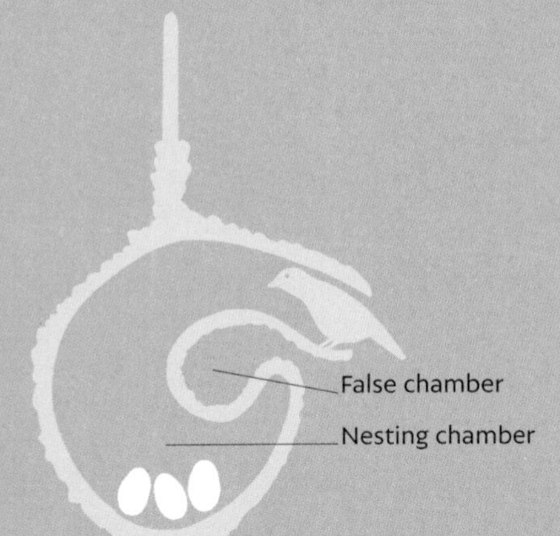

TERMITE MOUNDS

Termite mounds, in contrast, are built to last. They take years to create but endure for centuries. The caatinga mounds in northeast Brazil, for example, are made up of 200 million mounds spread over an area the size of Great Britain. They are thought to be between 690 and 3,800 years old. Unlike the tits' nests, which are designed to house a single brood, termite mounds are the ultimate in multigenerational social housing. They can be up to 30 feet (9 m) high. Built from a judicious mix of soil, saliva, and dung, which dries as hard as cement, they contain a network of ventilation shafts and a multichambered underground nest.

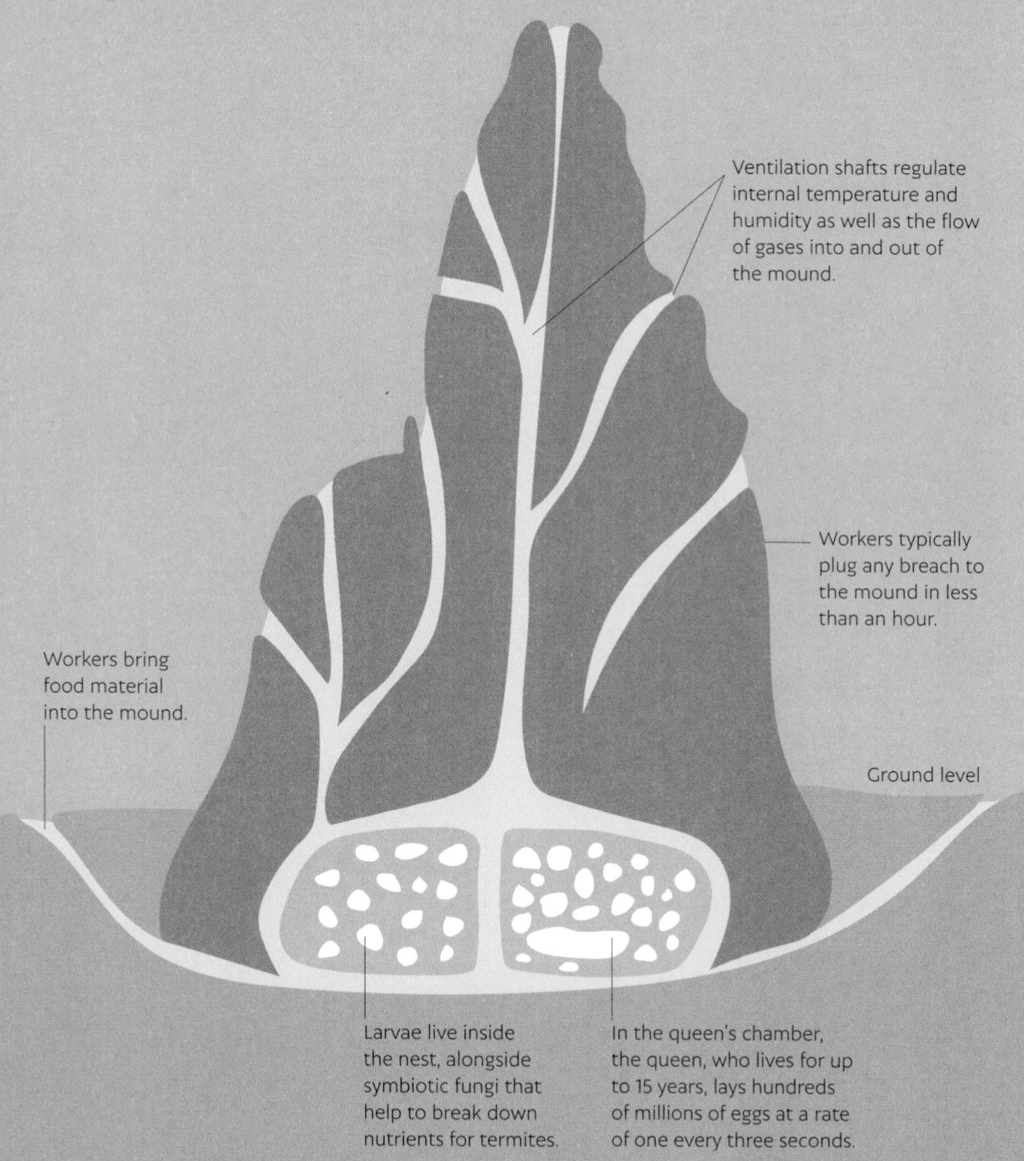

COURTSHIP

When it comes to courtship rituals, birds really know how to shake their tail feathers. Case in point are birds of paradise, a group of tropical birds famed for their elaborate mating dances. During the mating season, males spend many hours each day showing off. They shimmy, bop, glide, and hop, fanning their feathers and cutting the most outlandish shapes. The best dancers get to mate. For the kakapo, a flightless parrot from New Zealand, it's all about the bass. Males build stages on rocky outcrops, then "boom"—a repetitive, bass-like noise—from dusk till dawn, for months at a time. Other courtship rituals are equally impressive. Meet the penguin that's a natural-born mimic and the bowerbird that knows all about geometry.

THE MIMIC: EMPEROR PENGUIN

Length of courtship: 6 weeks

At the start of the Antarctic winter, thousands of emperor penguins leave their ocean feeding grounds and march hundreds of miles inland to their icy breeding area. The birds take a new mate each year. To attract a female, males wander through the colony making repetitive calls, each only a few seconds long. The pair then perform a graceful mimicking ritual, in which they synchronize their movements. Bowing, swaying, preening, even scratching are all performed in unison. With a pair-bond established, they then mate and waddle around until the female lays her single egg one month later. After this, she disappears to sea to feed. The female is gone for two months, during which the male incubates the egg; then they switch over. It is thought the courtship ritual aids bonding and helps the birds to recognize each other after their long periods of separation.

The mating dance of the emperor penguin

The illusory bower of the great bowerbird

THE ILLUSIONIST: BOWERBIRD

Length of courtship: Up to 2 months

Male bowerbirds spend months creating elaborate structures called "bowers" to impress potential mates. Bowers are made from natural products, such as sticks, feathers, and berries, with occasional human-made items, such as coins, nails, and bottle tops. Each item is placed with extreme precision. The great bowerbird of northern Australia even incorporates an optical illusion. Its stick tunnel features pebbles laid out in increasing size order, from front to back, creating the effect of forced perspective. From the female's position at the front of the tunnel, all the pebbles appear the same size. As a result, she may perceive the area to be smaller than it is, and the male to be bigger. It is a cunning tactic. Studies have shown that males who build the most geometric patterns are also the most likely to mate.

ANIMAL MAGIC

Elsewhere in the animal kingdom, practices are equally eclectic and bizarre. The male six-plumed bird of paradise, for example, puts on a delicate ballerina dance, while male *Drosophila* flies perform a detailed dance routine that involves wing-scissoring and rapid side-to-side steps.

Courtship rituals in the animal kingdom

Length of courtship

0 days

THE OPPORTUNIST

ASIAN CORN BORER MOTH

Asian corn borer moths have a clever way of avoiding predators. When they hear the ultrasonic call of a bat, they simply freeze and fall to the ground. When they want to mate, male moths use this quirk to their advantage. They imitate the bat's cry, causing the females to drop from the sky. While she is temporarily paralyzed, the male makes his move.

2–6 days

THE DUNG FLINGER

HIPPOPOTAMUS

Hippos are polygamous, meaning that males mate with multiple females. Dominant males get first choice and often spend days picking a mate. It's a testosterone-fueled time period, and fights between males often break out. To gain a female's attention, males urinate and defecate at the same time, then spray their would-be conquest with the heady cologne. It's a winning strategy, and shortly afterward, mating occurs.

It is not all about dance moves, though. Some animals adopt very different tactics to impress members of the opposite sex. Meet the opportunist moth, the dung-flinging hippopotamus, the artistic fish, and the echidna that likes to conga.

7–9 days

THE ARTIST

WHITE-SPOTTED PUFFERFISH

Male white-spotted pufferfish from Japan flap their fins to create ornate circular patterns on the seafloor, which they often decorate with fragments of shell. A little 5-inch (12 cm) long fish is all it takes to create an underwater artwork up to 6.5 feet (2 m) in diameter. Studies suggest that females may be able to use the structure's features to predict the size of its creator, helping them to pick the fittest mate.

30 days

THE STALKER

SHORT-BEAKED ECHIDNAS

Before courtship begins, the testicles of male short-beaked echidnas increase more than three-fold in size. The usually solitary males locate and then follow females, forming "trains" up to 11 males long. This bizarre conga can continue for a month, until the female stops and plants her head into the base of a nearby tree or bush. The males dig a doughnut-shaped trench around her, and push each other out of the way, until one gains the right to mate.

MATING MARATHONS

Courtship is one thing. Now, let's talk about sex. For some, it's over in seconds, yet for others, intercourse can last for hours, days, or even longer. It is also an evolutionary conundrum. Mating uses time and energy and leaves animals vulnerable to predation, so why do some animals undergo such marathon sex sessions? You would expect smaller animals, which starve quickly and are more likely to get eaten, to spend less time copulating than

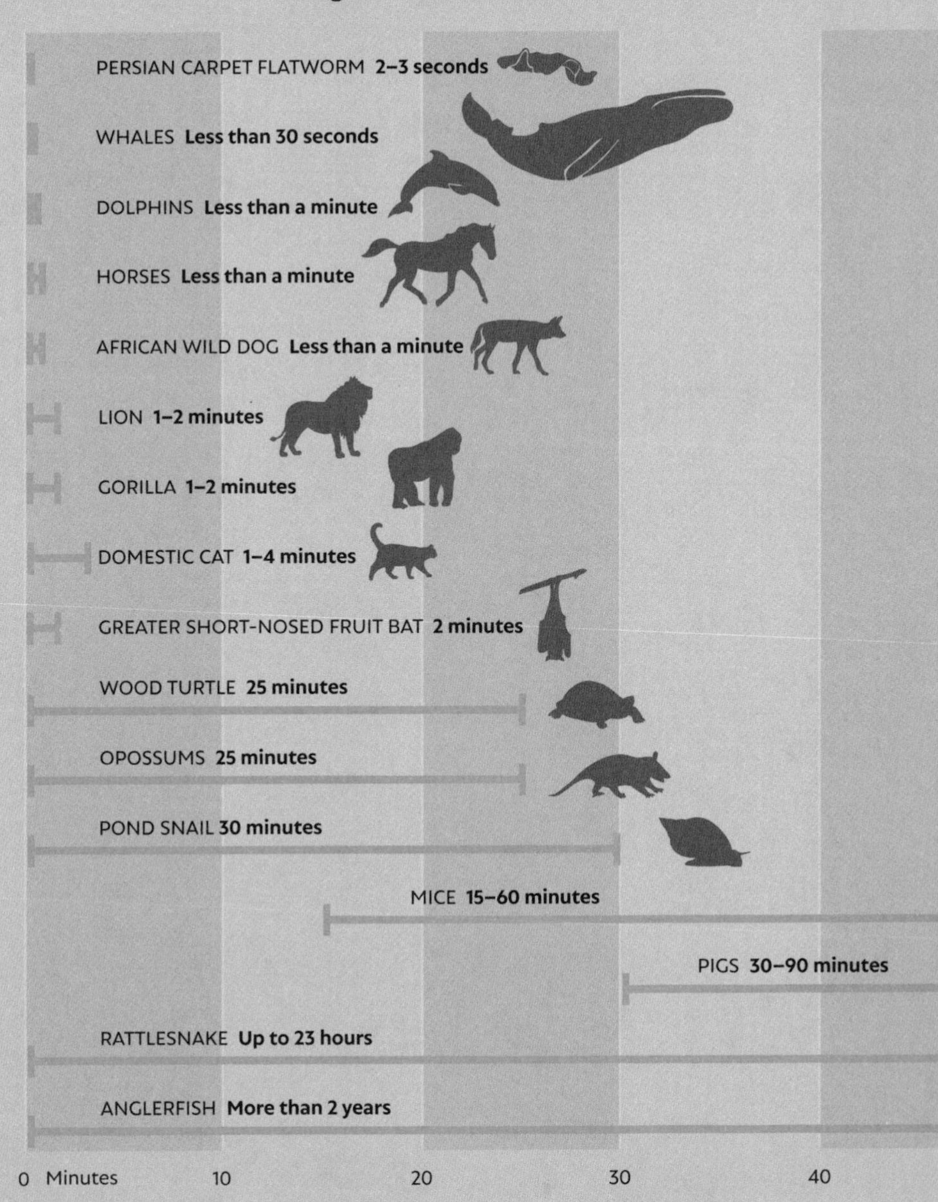

larger ones. Not so! A study of 113 mammalian species found that, in general, smaller mammals spend longer mating than larger ones. According to the scientists, this may be because small mammals are more capable of the "sustained energetic maneuvers involved in copulation." It could also be that prolonged mating prevents other males from accessing the female, and so helps to ensure paternity.

ORAL SEX IN BATS

Greater short-nosed fruit bats, from south and southeastern Asia, use oral sex to prolong the length of their otherwise brief liaisons. The practice buys the couple an extra 100 seconds of sex over the usual 2 minutes. There are various possible reasons for this. It may boost fertilization, make it easier for sperm to reach the oviduct, or give females the chance to pick up on chemical cues that help her to determine the suitability of her mate.

60 70 80 90

WEIRD SEX

ANGLERFISH

LENGTH OF INTERCOURSE: Years

When a male deep-sea anglerfish (*Linophryne indica*) finds a female, he bites into her belly and latches on until his body fuses with hers (see bottom right of image). Their blood vessels join, enabling the male to survive on the nutrients he receives from her bloodstream. In return, he provides her with sperm. The body parts he no longer uses, such as eyes, fins, and some internal organs, wither away, until he's little more than a sperm-filled bag. It is a macabre sex marathon that can last for years and is thought to have evolved because it gives the females, who live in an environment where few males are available, sperm "on tap."

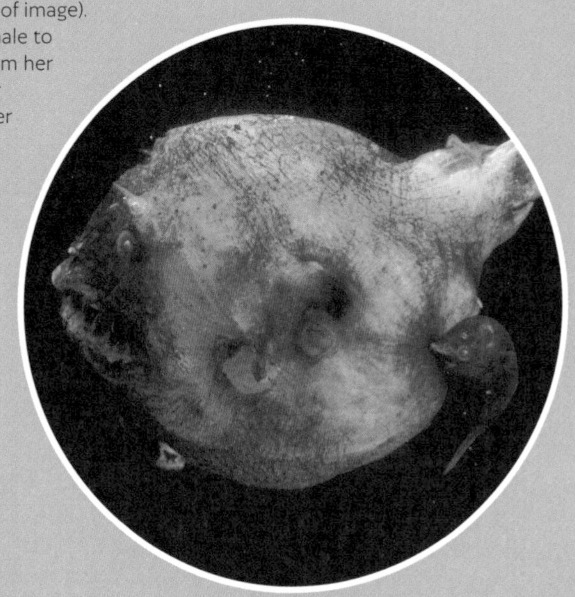

PERSIAN CARPET FLATWORM

LENGTH OF INTERCOURSE: 2 seconds

Persian carpet flatworms are hermaphrodites, producing both sperm and eggs. To avoid the labor of looking after fertilized eggs, individuals try to make others pregnant, while avoiding becoming pregnant themselves. This is achieved by penis fencing. Duos spend up to an hour jousting with their sharp, two-pronged penises, trying to stab and inseminate each other. When it happens, insemination takes only a few seconds. The victor then becomes the "father" and the stabbed individual becomes the "mother."

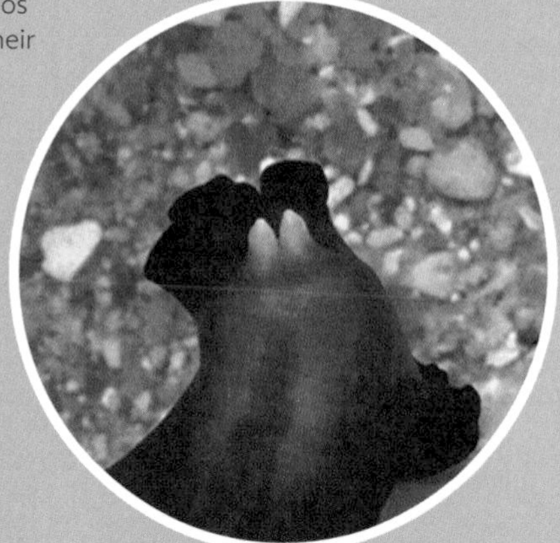

DYING OF SEX

Sexual intercourse leads to the creation of new life, but for some animals, it also brings about their demise. Semelparity, also known as suicidal reproduction, occurs when an animal concentrates all of its energy into a single bout of mating, then dies either during or just after sex.

BROWN ANTECHINUS

LENGTH OF INTERCOURSE: **Several hours**

Sex can be a risky business for mammals, too. The brown antechinus is a nocturnal marsupial mouse from Australia. As soon as they become sexually mature, at 10 months old, males enter into a frenzied 2-week-long sex marathon. During this time, males mate with as many females as possible. Sex lasts for several hours or more, and as soon as he is done, the male moves on to find another partner. As a result, he has no time to eat, drink, or sleep. It's exhausting. As his levels of stress hormones rocket, his immune system crashes. Males become prone to parasitic infections and gastrointestinal ulcers. They die shortly afterward and never live to see a second mating season.

ORB WEAVER SPIDER

LENGTH OF INTERCOURSE: **20 seconds**

Some spider species practice suicidal reproduction. Male spiders have a pair of armlike appendages called "pedipalps" that they use to transfer sperm into a female's body. For the male orb weaver spider, the process takes 20 seconds, but he dies as he inserts his second pedipalp into her. His body is left hanging from the female's genital opening, which prevents other males from mating with her. Often, she will eat him in a practice called "sexual cannibalism." Such postcoital snacking is common among insects and arachnids but has also been seen in some snails and copepods (tiny aquatic crustaceans).

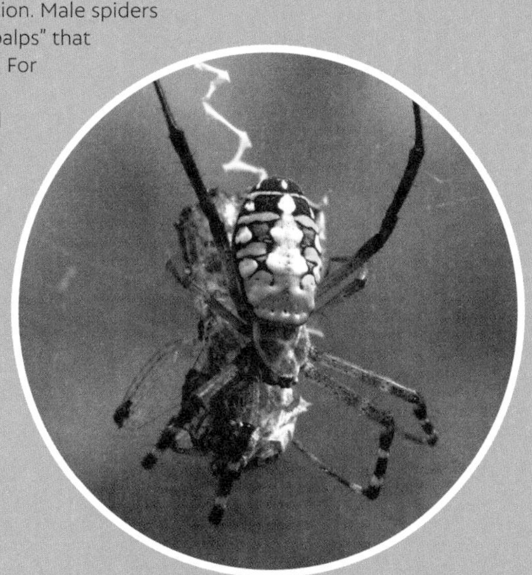

It doesn't always end well for females, however. For example, the female desert spider has sex and then feeds her newly hatched spiderlings with regurgitated food. After that, she performs the ultimate sacrifice and lets her spiderlings eat her.

SEASONAL SHIFTS

Warmer global temperatures are causing the seasons to shift. Spring is arriving earlier, and the winters are shorter. There are more hot days, fewer freezing days, and unseasonal weather events are becoming more common. The oceans are warming. From fishing to flowering to hibernation, this is changing the way that animals and plants behave. Organisms able to adapt their behavior to match the shifting seasons may have a shot at survival, but for those who are more fixed in their ways, the future is less certain.

The effect of climate change on puffins and their food source

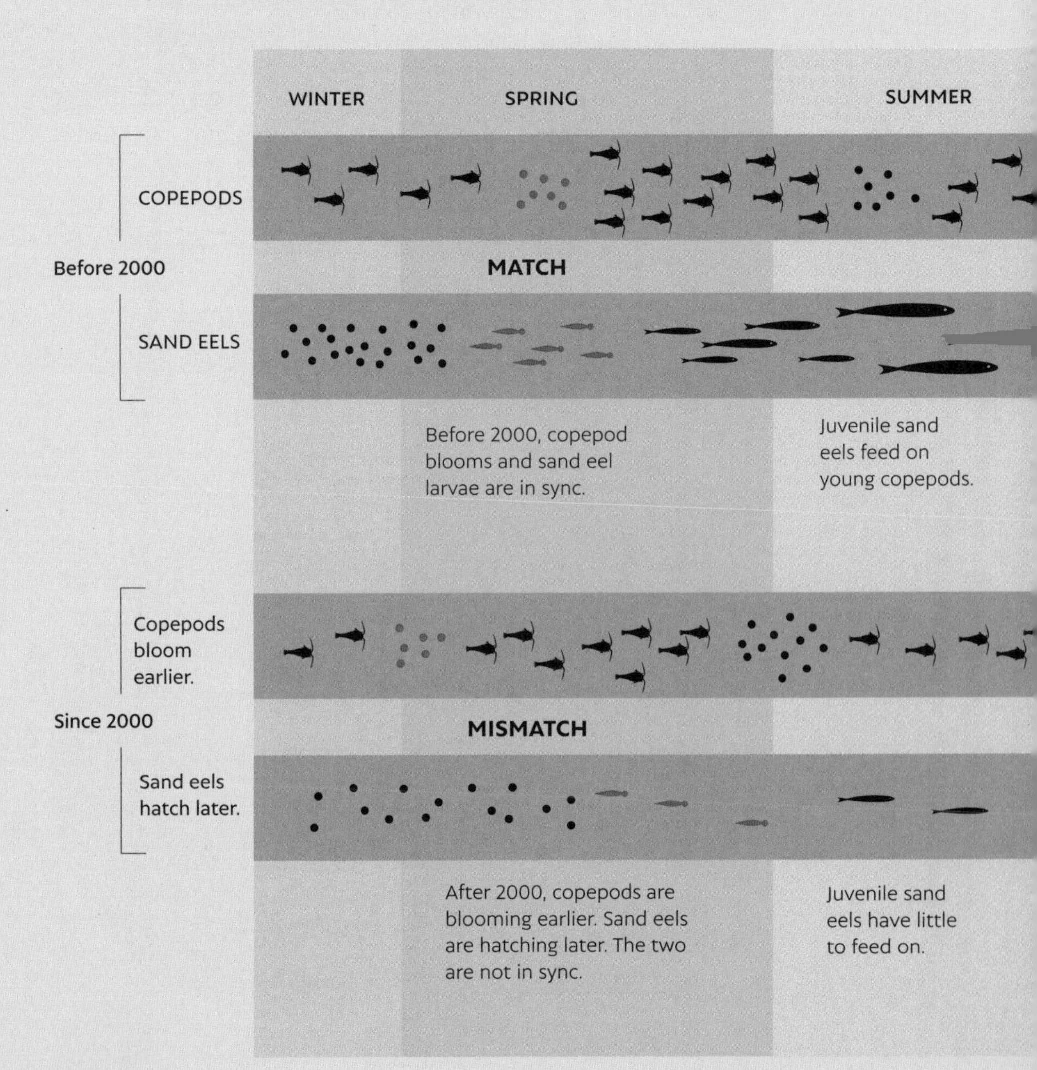

PUFFINS

Puffins are a case in point. These charismatic seabirds eat sand eels, which eat tiny crustaceans called "copepods." In days gone by, vast "blooms" of young copepods coincided with the hatching of sand eel larvae, which then grew and were eaten by puffins, who fed the sand eels to their chicks. These days, however, warming waters and shifting seasons are causing the food chain to be out of sync. Copepods are now blooming around three weeks before the sand eels hatch. As a result, there are fewer adult sand eels available for puffins to feed to their young. Adults are spending more time fishing, often to no avail. Their chicks are starving. In Norway's Røst archipelago, puffins have declined by more than 80 percent in the last 50 years.

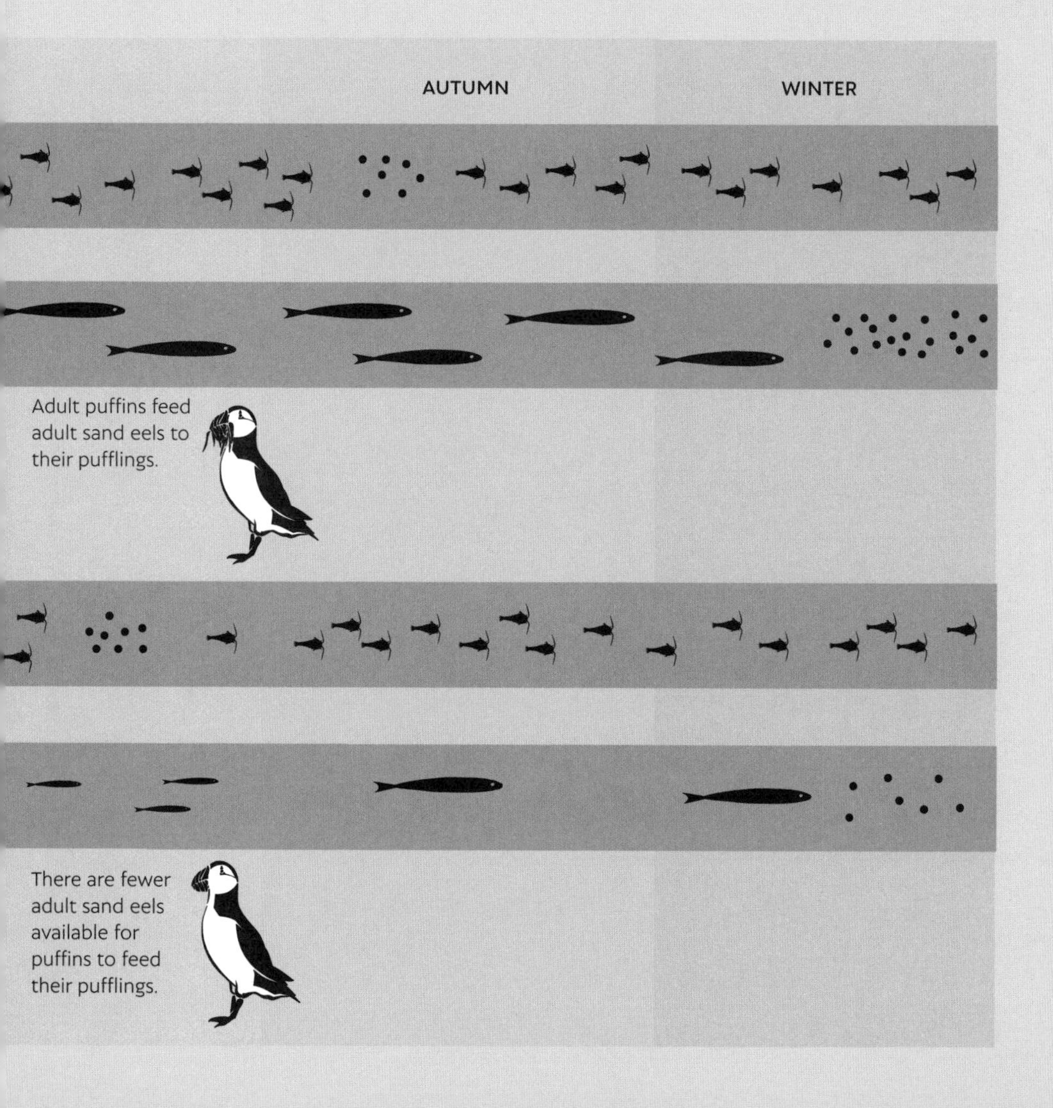

BIOLOGICAL SPANS

INTRODUCTION

All living things are made from cells—tiny, complicated bags of "stuff" that contain both the instructions and machinery needed to orchestrate life. When cells combine, they acquire a higher purpose. They become tissues, such as the nervous tissue that fires electrical impulses around the body and the muscle tissue that helps our hearts to beat. Working together, these tissues become systems: the nervous system that coordinates thought, or the cardiac system that pumps oxygen-rich blood around our bodies. When these systems are viewed as a whole, a complex organism emerges, replete with its own internal clocks.

These clocks generate rhythms that operate on time spans from minutes to hours to days to years. Think of the beating of your heart or the rise and fall of your breath. Internal rhythms determine when an organism's reproductive life begins and ends. They also determine the pattern of fertility that falls somewhere in the middle. Rabbits, for example, have an estrous cycle that lasts for about 12 to 16 days, and they are fertile for most of this period. Pandas, in sharp contrast, are only fertile for between one and three days, at some point between February and May.

The rhythms of life are influenced by internal factors, such as the cyclical patterns of genetic activity or the rise and fall of hormones. However, they are also shaped by external factors. Corals and some marine worms time their spawning to the phases of the moon, while the morphing of the seasons and the rising of the sun influence everything from plant growth and flowering to the mass migration of some animals.

Biology influences time spans in other ways, too. Sound is just a wave, traveling through time, but the frequency and shape of that wave is constrained by biology. From the deep, infrasonic rumbles of the Sumatran rhino to the high-pitched chirping of a locust, animal vocalizations are influenced by internal anatomy.

Behavior is influenced by internal factors, such as hormones, and external factors, such as the waxing and waning of the moon.

ESTROUS CYCLES

How can you tell if a giant panda is pregnant? With over 500 of the bamboo-munchers currently in captivity, and with their keepers all keen to help the global panda population grow, it is a question of great importance. The answer comes from a thorough understanding of the panda's estrous cycle.

Estrous cycles provide female mammals with repeated opportunities to become pregnant. They are the recurring physiological changes that occur over time in response to fluctuating hormone levels, and they differ greatly between species.

Some species, such as cats, humans, and cows, are polyestrous, meaning they go through multiple cycles per year. Women have a cycle that lasts around 4 weeks, cats have a cycle of 2 to 3 weeks, elephants have a cycle of 13 to 18 weeks, while hamsters have a consistent estrous cycle that lasts only 4 days. That's why it's possible to go from two hamsters to too many hamsters in a very short amount of time.

Monestrous species, such as wolves and bears, however, only have one cycle per year. Female giant pandas are only fertile for 24 to 72 hours, once every 12 months, at some

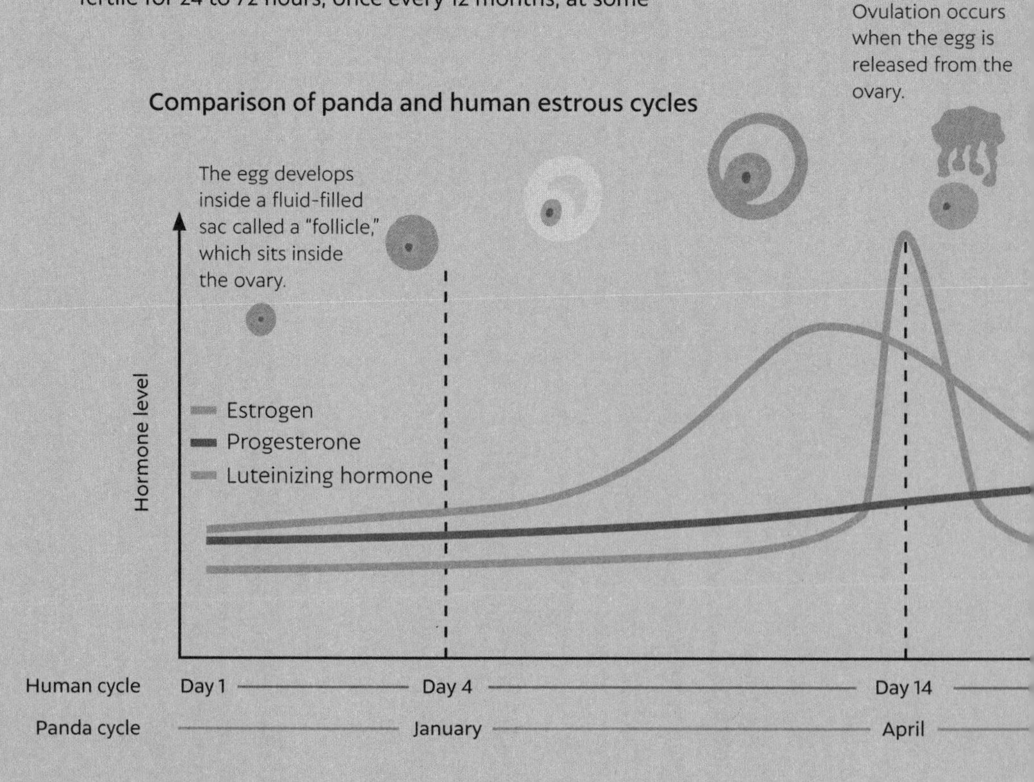

point between February and May. It's when the bamboo is at its sweetest, but blink and you'll miss it.

The only way to anticipate this cycle is by regular urine tests. Levels of two sex hormones, estrogen and luteinizing hormone, rise as ovulation approaches. After they peak, it is time to either let the female panda breed naturally or use artificial insemination in captivity.

After that, it's not easy to tell if the panda is pregnant—a fertilized egg can sometimes sit around in the womb for 6–12 weeks before eventually implanting into the lining of the womb and starting to develop. Sometimes, inexplicably, developing embryos are reabsorbed by the mother, and sometimes, pseudopregnancies occur—hormonal and behavioral changes happen, but there is no panda cub.

Genuine pregnancy is best detected by measuring the relative levels of luteinizing hormone and progesterone, and by ultrasound, which is possible in the last 20 days of the panda's 140-day gestation. Even then, pandas are full of surprises. Sometimes, the first a keeper knows of a panda pregnancy is when the panda cub arrives.

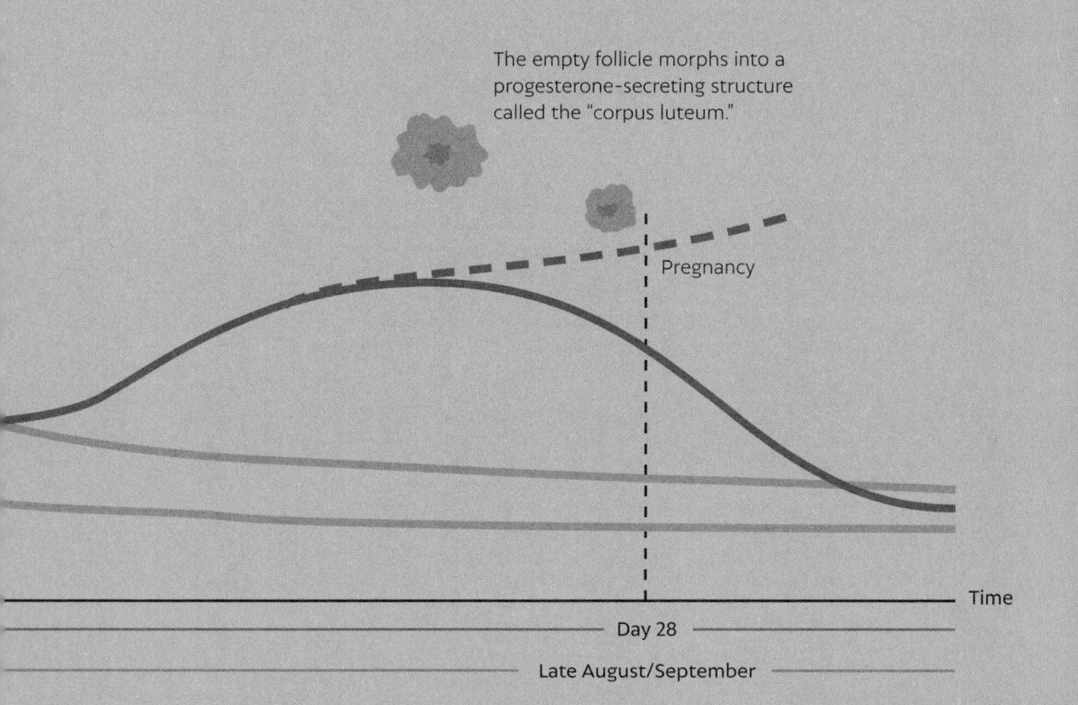

REPRODUCTIVE SPANS

Menopause is an oddity. Most female mammals remain fertile throughout their lives, then die about the same time as their ovaries stop functioning. Female humans, on the other hand, can live for decades after their periods stop—and they are not alone.

We share menopause, not with other primate species but with a handful of marine mammals. It is well known that adult killer whales and short-finned pilot whales experience menopause, but it's recently been discovered that narwhals and belugas also live well into their non-fertile years.

A 2018 study looked at 16 different species of toothed whales. Scientists studied teeth, which give an indication of age, and ovaries, which give an indication of fertility. From this, they could work out how long each whale lived after she stopped being fertile. Most whale species were fertile right up until they died, but narwhals, belugas, and short-finned pilot whales had a much longer post-reproductive span than other species. This suggests that they go through menopause the same as humans.

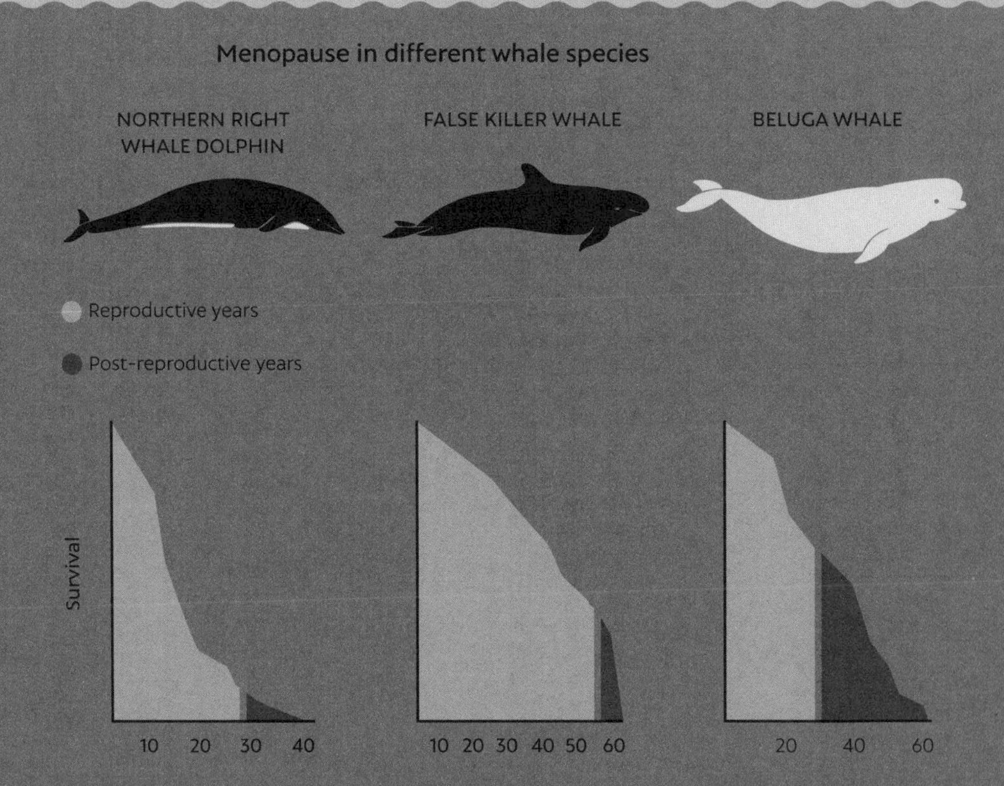

The ability to live through a non-fertile period of life has evolved at least three separate times in toothed whales: once for killer whales, once for short-finned pilot whales, and once for the common ancestor of narwhals and belugas. The question is why? If evolution favors strategies that help organisms to reproduce and pass on their genes, why squander resources on a time period when offspring can no longer be produced?

The answer may lie in the "grandmother effect." Killer whales live in large, matrilineal pods, and grandmother whales have been spotted sharing fish with their grand-calves. They may also teach the younger generations where to find food. So, when grandmothers help to look after the younger generations, they make it more likely that their genetic line will continue.

It is easy to see how the same could hold true for our own species, where the survival of our ancestors was probably heavily dependent on intergenerational cooperation. As for narwhals, belugas, and short-finned pilot whales, they all have different social structures, so the mystery of menopause remains.

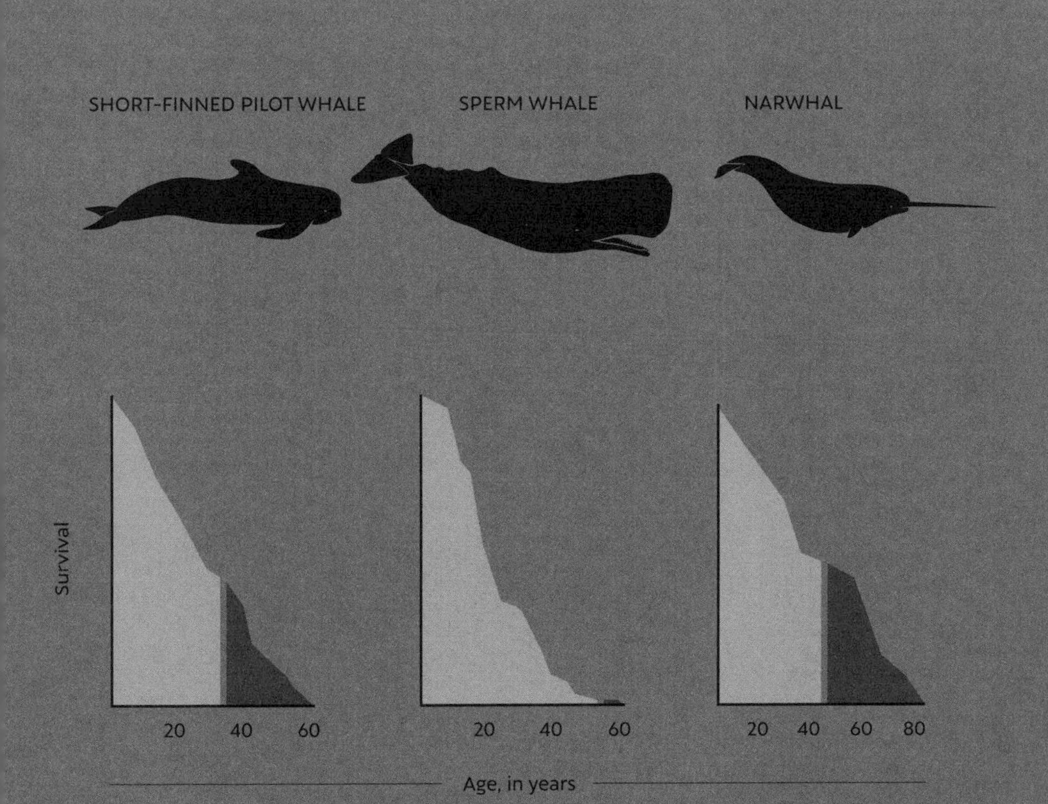

RHYTHMS OF LIFE

How do birds know when it is time to wake up? How do flowers know when to open and close? And how do monarch butterflies manage their annual migration?

These rhythmic patterns of behavior and biology are all controlled by biological clocks. Most animals, plants, fungi, and bacteria possess an inner clock. They respond to external factors, like temperature and day length, and internal factors, such as changes in gene activity, to influence the vital processes of feeding, sleeping, and reproduction.

Biological clocks provide an internal assessment of the passing of time, and in some cases, this helps with navigation. Every year, monarch butterflies fly thousands of miles from the US and Canada, where they breed, to the forests of central Mexico to hibernate.

Monarch migration

Monarch butterflies use a circadian clock inside their antennae to keep track of time as the sun crosses the sky. This enables them to compensate for the sun's movement and maintain a constant southward heading.

The butterflies head south, using the sun to navigate, but as the sun moves across the sky, they have to recalibrate their bearings to keep on track. To do this, they have to be aware of time, which is provided by their body clock.

Where the underlying biology oscillates on a 24-hour cycle, such as for the monarch butterfly, it's called a "circadian clock." Scientists now have a good idea of the cells and molecules that are involved. In vertebrates, the clock resides in a cluster of around 20,000 neurons, called the "suprachiasmatic nucleus," inside the brain, but in butterflies, it's found in the antennae. If an insect's antenna is clipped or covered up, it becomes disoriented and unable to pursue its southward heading.

If the antennae are covered and the clock stops functioning, the butterflies are unable to tell the time or navigate accurately. They simply follow the sun and can veer dramatically off course.

LUNAR CYCLES

While some behaviors, such as foraging and flowering, need to be regulated on a daily basis, others are managed over much longer time periods. Circannual cycles run over a period of one year, while lunar cycles operate in sync with the waxing and waning of the moon.

In some marine organisms, mating is linked to the stages of the tide, which are controlled by the phases of the moon. Every year, millions of red crabs march hundreds of miles from Christmas Island's inland forests to its beaches, to mate and spawn.

Red crab located on Christmas Island

The life cycle of the Samoan palolo worm

The palolo worm has a remarkable life cycle. When the moon is in its third quarter, the worms rise up and spawn for a few hours at a time, across three nights, always starting at around 2 AM.

Tiny free-swimming larvae drift to the ocean floor.

This synchronous spawning puts eggs and sperm in close proximity to one another. Fertilization occurs.

Larvae develop into adults and bury themselves into the reef.

As the neap tide approaches, the worm's tail end changes. Muscles and organs degenerate, and reproductive organs swell.

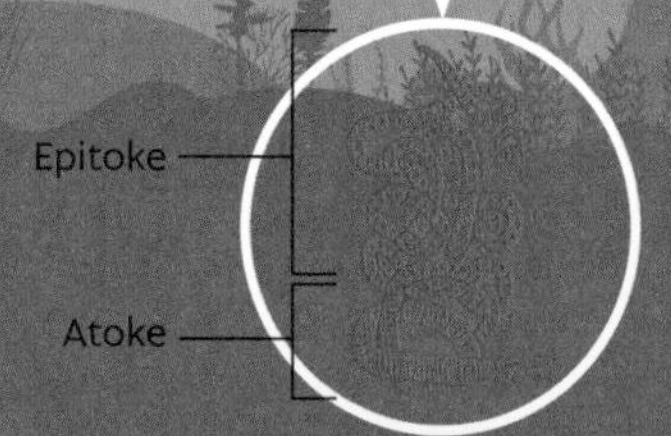

Spawning always occurs before dawn on a receding high tide during the last quarter of the moon. Remarkably, the crustaceans know exactly when to leave their forest burrows in order to make this date.

The Samoan palolo worm is equally precise with its timing. This segmented worm spawns only during the third quarter of the moon, in October or November. The event is eagerly anticipated by locals, who scoop up and eat the sticky worms. Fried with butter and onions, they are said to taste like a cross between oysters and mussels.

Epitokes that are not scooped up and eaten burst to create a thick, gloopy soup made of eggs and sperm.

Millions of epitokes are released at the same time and swim to the surface.

The worms stick their tail ends out of the rock.

When the lunar timing is just right, the tail end breaks off to form a separate animal called an "epitoke." The head end, known as the "atoke," remains attached to the seafloor and regrows a new tail.

WHAT GOES IN . . .

What goes in must come out. Food enters the body via the mouth, and "leftovers" leave via the anus. The time spent "in the middle" varies greatly. In humans, the time from mouth to anus, known as "bowel transit time," is measured by getting people to swallow a special pill with a wireless transmitter.

Nonhuman animals are less keen on this method, so scientists instead use a measure called "mean retention time" (MRT), which is the average time a marker remains in the gut. These markers can be foods, such as pellets or leaves, or small, nondigestible plastic beads that pass straight through the body.

Generally, cold-blooded animals take longer to digest their food than warm-blooded animals, bigger animals take longer than smaller ones, herbivores take longer than carnivores, and mammals take longer than birds. Exceptions exist, and the time taken to process food depends on many factors, including anatomy, physiology, and the amount and type of food consumed.

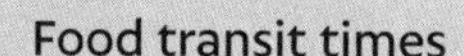

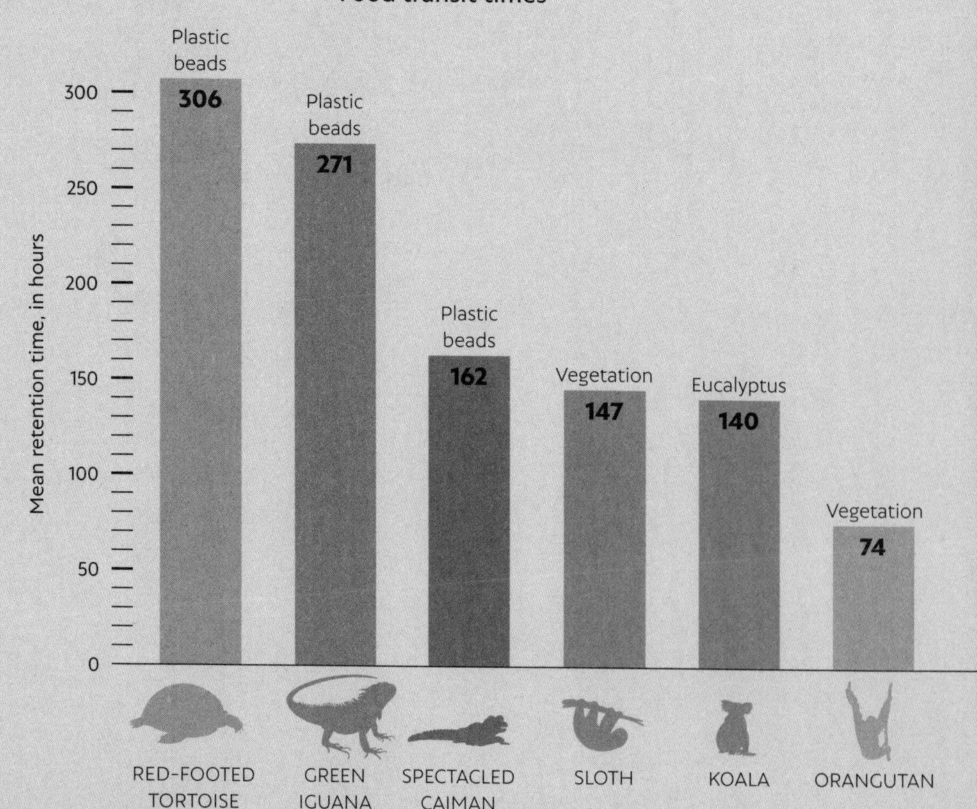

SLOTHS, MOTHS, AND ALGAE

Sloths only defecate once a week, and they lose up to one-third of their body weight in a single stool. Unlike other tree-dwelling mammals, such as monkeys, which defecate from the canopy, sloths clamber down to the ground and always go in the same spot at the base of the tree. It takes a lot of energy and makes them vulnerable to predators, so why do it?

Some think the concentrated pile of dung helps solitary sloths to find one another; others believe the animals are fertilizing their favorite tree. Another suggestion centers on the relationship that exists between sloths, moths, and the algae that live on them. Algae, which give the sloth its greenish tinge, comprise about 3 percent of the sloth's biomass. Sloth, moth, and algae are all part of a cycle that sees moths provide food for the algae, the algae provide nutrients for the sloth, and the sloth provide a home and free transport for the moth. When the sloth defecates on the ground, it provides a vital resource that enables the moth to complete its life cycle.

When the moths die on the sloth's fur, they release nutrients to the algae, which are then eaten by the sloth. It provides a much needed energy boost to an otherwise lackluster diet.

Sloth–moth relationship

Freshly hatched moths fly up to the sloth and colonize its fur.

The moths live, mate, and die on the sloth.

Female moths hitch a ride to the forest floor and lay their eggs in the sloth's feces. Caterpillars eat the poop and pupate on the ground.

TIME TO GO

Nearly all animals defecate. These droppings vary greatly in both size and shape. Wombats produce cube-shaped feces; fox poop is long and twisted; blue whales can excrete up to 400 gallons (200 L) of feces in a single bowel movement.

However, land-living mammals, regardless of size, take roughly the same time to "go." In a 2017 study, researchers filmed elephants, giant pandas, and warthogs defecating at a local zoo as well as one of their dogs in the park. They also studied 19 videos of different mammals from YouTube. Just like humans, all of the animals produced cylindrical feces, yet despite their various sizes, they all took about the same time to empty their bowels.

It's a joyful finding that unites the mammalian kingdom, and is made possible by a few key features. Firstly, it doesn't matter if the animal is big or small; the normal, low-level amount of pressure used to expel feces is constant. We all "push" with a similar degree of force. Secondly, feces slide along the large

Average times for animals to defecate

Researchers studied 23 different mammals, ranging in mass from a 9-pound (4 kg) cat to a 11,000-pound (5,000 kg) elephant, and recorded the time they took to defecate. The study showed that despite the large variation in mass, each animal took around 12 seconds to poop.

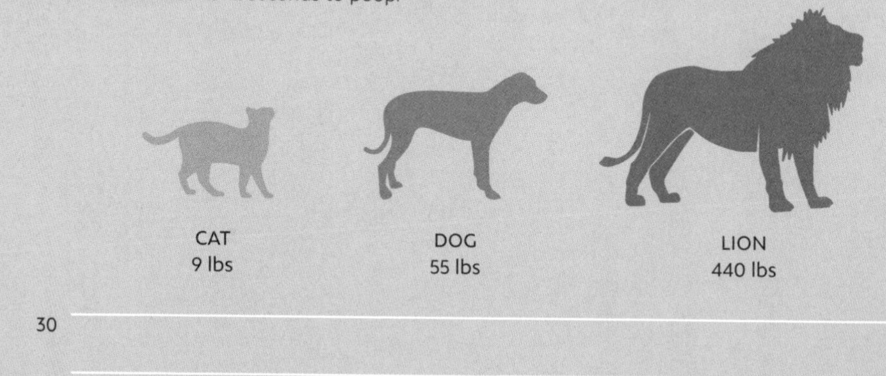

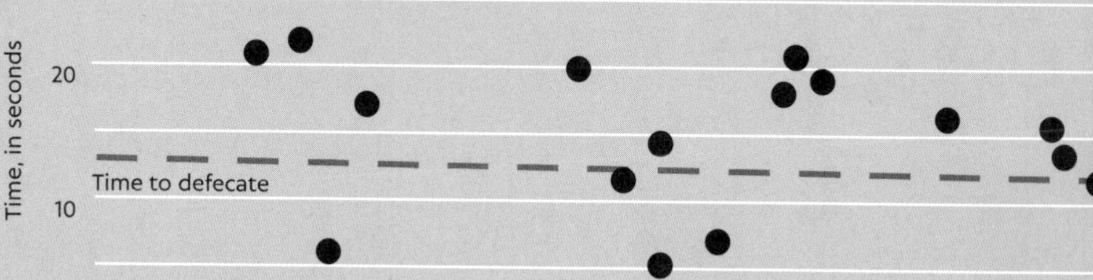

intestine on a layer of mucus. Bigger animals may have longer feces and longer rectums, but they also have thicker mucus. This makes the feces accelerate more, so they can cover greater distances in the same amount of time.

The same group of scientists also studied how long it takes mammals to urinate, enabling them to formulate the lofty "law of urination." Relying on yet more YouTube footage, they amassed data on body mass, bladder pressure, and the length of the urethra (the tube that connects the bladder to the outside world), and found that medium- to large-size mammals all take around 21 seconds to empty their full bladders.

The average elephant, for example, has 3-foot-long (1 m) urethra that is as wide as a drainpipe. This helps the urine to accelerate as it travels along, so although there is more urine, it still exits in the typical time frame. Medium-size animals, like dogs, meanwhile, have shorter urethras, slower flows, and smaller bladders, so it all balances out.

● Each dot represents a data point from an individual animal.

– – The average time that different-size animals take to defecate

RHINO
4,400 lbs

ELEPHANT
11,000 lbs

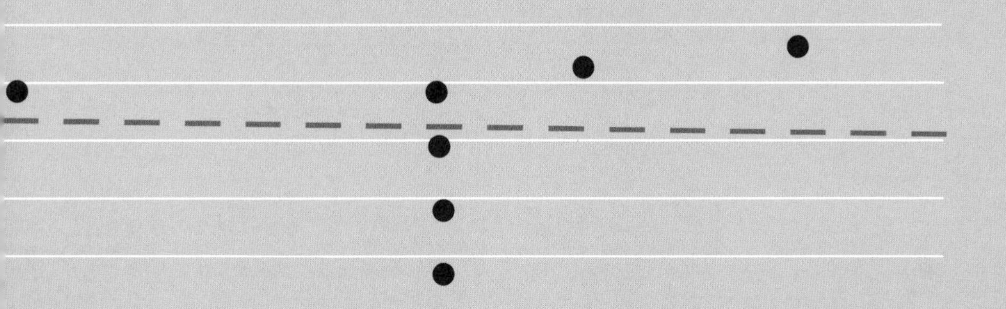

BREATHE DEEP

On March 27, 2021, Budimir Šobat from Croatia bagged a Guinness World Record when he held his breath for a remarkable 24 minutes and 37 seconds. It's the longest voluntary "breath hold" by a human, but it pales into insignificance when you discover the air-gulping abilities of some other mammals.

On their foraging trips for squid and fish, beaked whales dive deeper and longer than any other mammal. A 2020 study of 3,680 dives made by 23 beaked whales found that half of all foraging dives lasted more than an hour. One in 20 exceeded 77 minutes. The record, however, goes to an individual off the southern California coast who, in 2014, stayed submerged for 2 hours and 17 minutes while diving to a depth of 9,816 feet (2,992 m). So, how do they do it?

When they begin a deep dive, whales hold their breath, slow their heart rate, and shunt blood away from their extremities to where it is needed most: the brain, heart, and muscles. Water pressure increases with depth. At around 656 feet (200 m), this causes the lungs to collapse, so the whale draws on oxygen stored in its blood and muscles. Whales have unusually high levels of 2 oxygen-carrying proteins: hemoglobin and myoglobin. Myoglobin, for example, is 10 times more concentrated in the muscles of diving mammals such as whales, seals, and beavers than it is in human muscles. This enhances the oxygen reserves of these specialist diving mammals and helps them breathe for extended periods underwater.

Breath-holding during dives by Cuvier's beaked whale

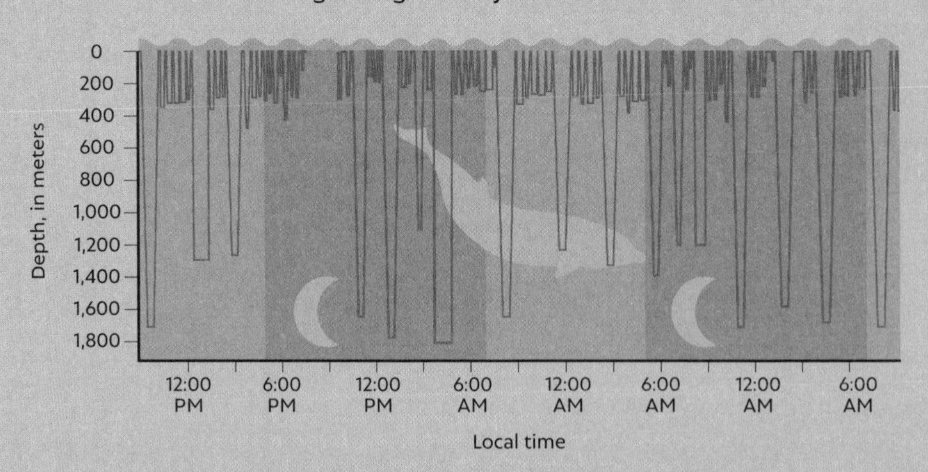

Cuvier's beaked whales alternate shorter, shallower dives, with longer, deeper ones. Most deep dives are to depths of around 3,937–5,905 feet (1,200–1,800 m). Half of these last for more than an hour.

BREATHE, SLEEP, SWIM

Marine mammals such as seals and whales may well be able to hold their breath, but how do they sleep without drowning? Bottlenose dolphins spend around one-third of their day asleep, but it's more of a nap than a sleep. The dolphin shuts down half of its brain and the opposite eye. The other side of its brain remains on "low alert," enabling the animal to keep watch for predators and detect when a breath is needed. Dolphins take 8–10 breaths per minute when they are awake, but this slows to 3–7 breaths per minute during rest. After two hours, the halves of the brain reverse roles, and the snoozing continues.

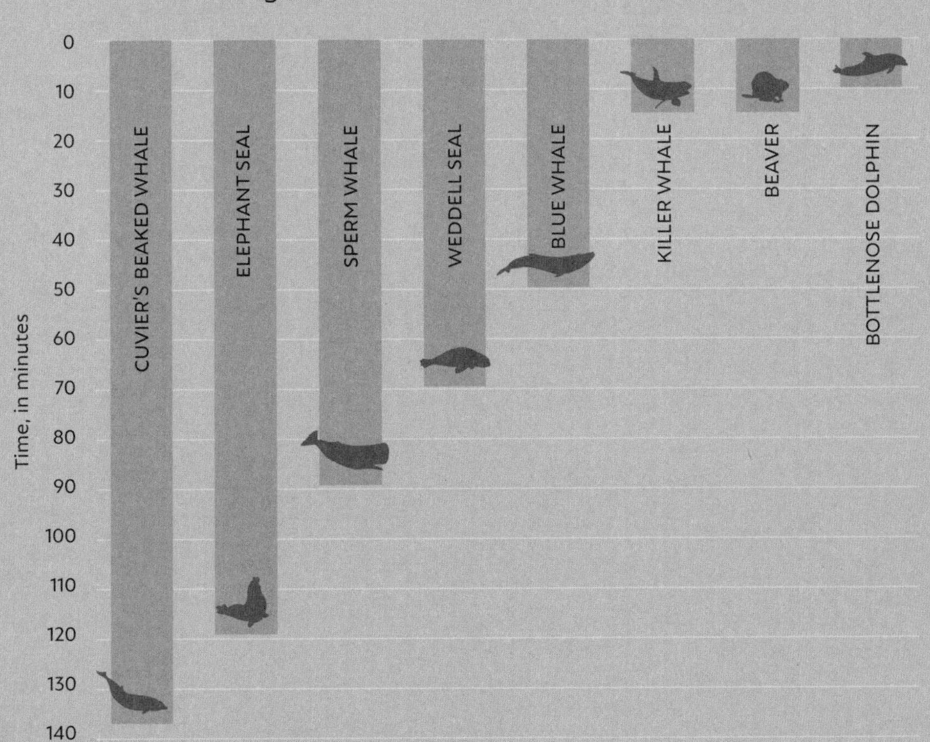

A FISH OUT OF WATER

With their scaly, streamlined bodies, fish are well adapted for a life in water. They obtain the oxygen they need to breathe as water flows into their mouths and over their gills. The gills have a large surface area and are rich in tiny, thin-walled blood vessels called "capillaries." Oxygen moves from the water into the fish's capillaries, and waste carbon dioxide moves in the opposite direction. It's an efficient process, but it relies on the fish being in water.

Take fish out of water, and most die within minutes. But this is not the case for the mangrove rivulus, a plucky little fish that can survive for up to 66 days on the land. It does this not by holding its breath but by switching from breathing via its gills to breathing through its skin.

The breath-holding ability of the mangrove rivulus fish

The mangrove rivulus lives in the mangrove forests of the southern United States and South and Central America.

When the brackish pools that it lives in get too hot, too salty, too crowded, or too dry, the fish uses its tail to flip itself out of the water and onto land.

Some other fish species can breathe out of water—although for shorter time periods. The northern snakehead fish, for example, is an invasive species to the US, where it is considered harmful to the environment. It has an unusual respiratory system, including specialized structures above the gills, which enable the fish to survive out of water for several days. During this time, it can wriggle its way into new streams and lakes, making it hard to contain.

As it wriggles around, looking for somewhere more hospitable to live, the fish's physiology begins to change. The first, rapid raft of changes happen within a day.

Within a day

The fish prepares to stop breathing through its gills. The activity of four key genes increases, triggering the growth of blood capillaries close to the surface of the skin.

Within a day, the fish is breathing through its skin. Oxygen passes from the air, through the skin into the newly formed capillaries. Despite inhaling less oxygen overall, the fish can still maintain its original metabolic rate.

During prolonged dry spells, dozens of fish hole up inside moist crab burrows and decaying logs. Over the next week, their physiology changes further.

Within a week

Within a week, the gills have stopped working altogether. A mass of cells grows inside the gills to prevent the delicate structures from collapsing. The fish continues to breathe through its skin, but now its metabolic rate is reduced. This enables the fish to eke out its dwindling energy reserves. Nitrogen and other waste products are excreted through the skin, rather than the gills.

EVERY BREATH YOU TAKE

Breathe in, breathe out, and repeat as necessary. Animals vary greatly in their respiratory rates: the number of breaths they take per minute. Respiratory rate is influenced by many factors, including size, physiology, and the level of activity. The tiny hummingbird takes 250–300 breaths per minute. Giraffes, the world's tallest mammal, take only seven breaths per minute, and alligators breathe even less frequently. Bears, meanwhile, usually inhale and exhale around six to eight times per minute, but this is massively reduced during hibernation. A slumbering wintry bear takes only one breath every 45 seconds.

How many breaths per minute?

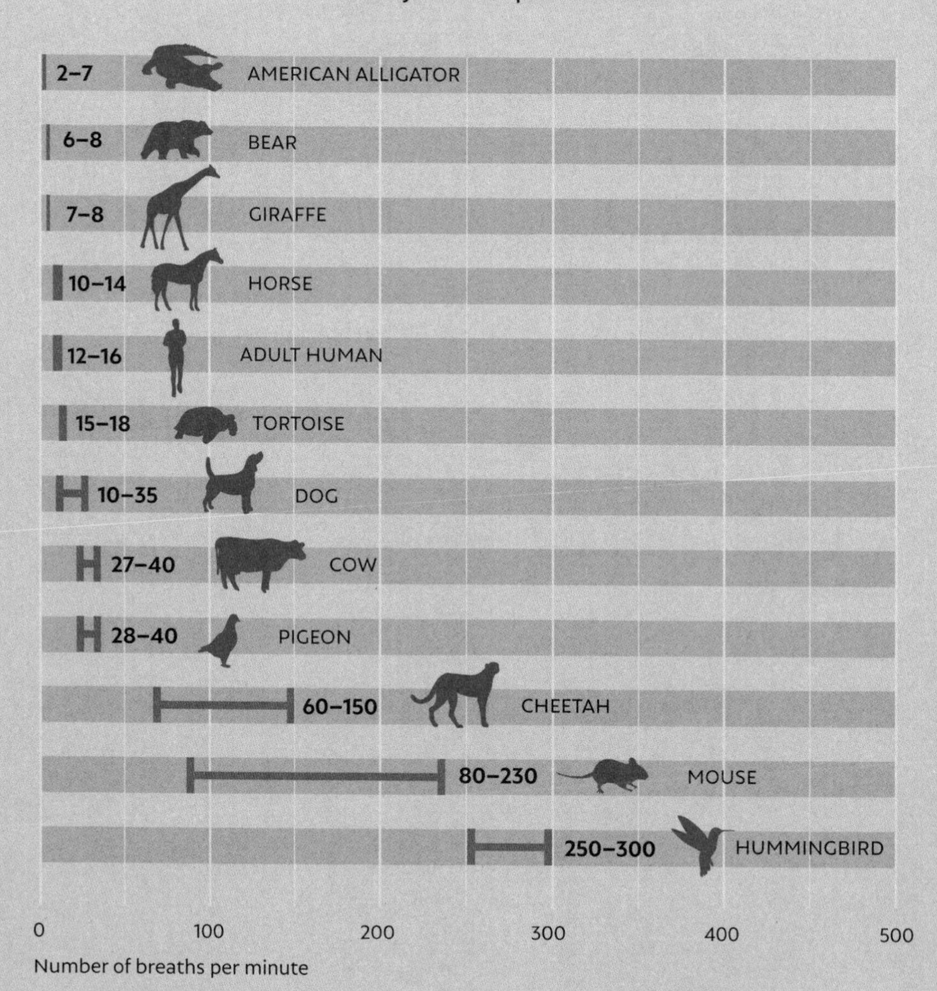

A BREATH OF FRESH AIR

The respiratory system of birds is very different to that of mammals. Birds don't have a diaphragm, but they do have lungs and a series of air sacs that they use like bellows to move air through their lungs.

It takes two full breaths for air to move in and out of the bird, but there is always a continuous flow of oxygenated air passing over the bird's lungs. This makes bird respiration very efficient, because oxygen is entering the bird's body both when it breathes in and when it breathes out.

This unidirectional air flow enables birds like the bar-headed goose to fly over the Himalayas even though the air is thin. These high-fliers have been recorded at altitudes of over 22,900 feet (7,000 m), and mountaineers say they have seen them flying over Mount Everest. Unidirectional air flow also enables bird to sing without pausing to take a breath.

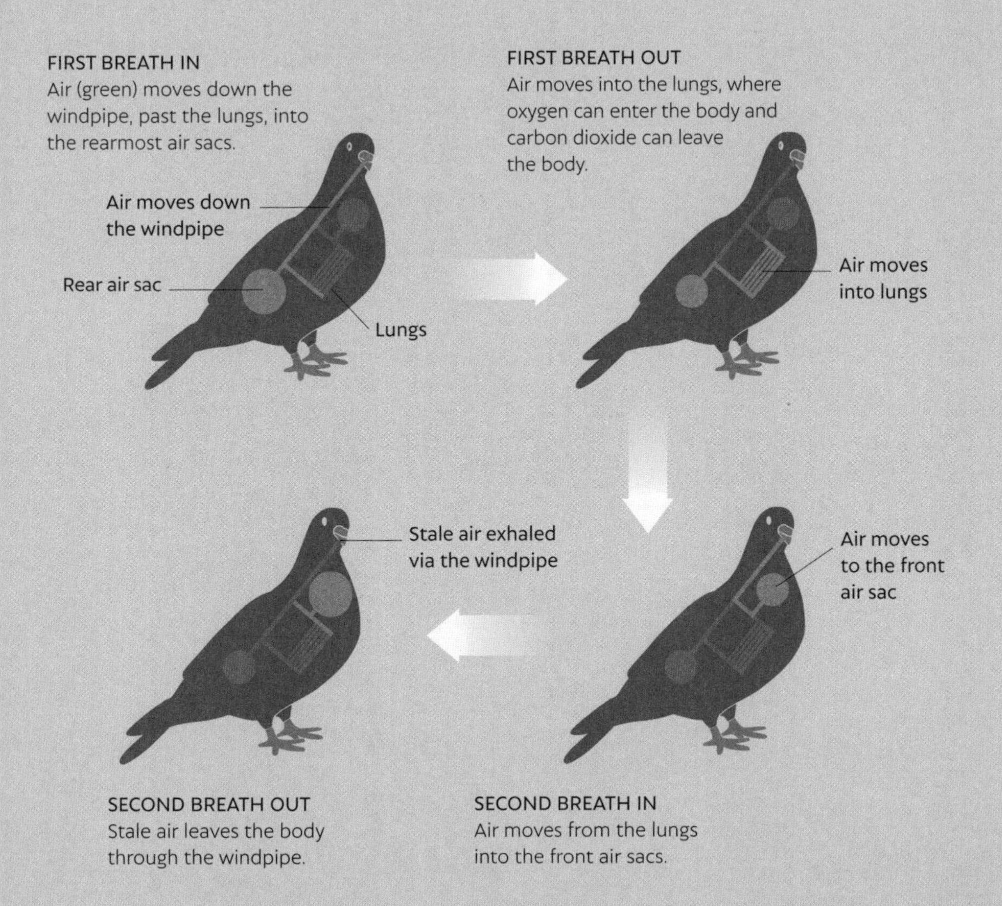

THE SPEED OF SOUND

In 1967, biologist Roger Payne lowered a microphone into the ocean and recorded the most incredible sound: the haunting song of the humpback whale. His recordings became an album and helped to leverage support for a future ban on commercial whaling. Sounds, it seems, can be both beautiful and powerful.

Sound travels in waves. The frequency of a sound is the number of waves produced by a source each second, and it is measured in hertz (Hz). Living things communicate with sound of varying frequencies. The melodies of the humpback, for example, fall between 20 Hz and 4,000 Hz.

Adult human speech occurs at a frequency of 85–255 Hz, but our range of hearing is much greater from 20–20,000 Hz. Any sound with a frequency below 16 Hz is called "infrasound" and any sound with a frequency above 20,000 Hz is known as "ultrasound." Both are inaudible to humans.

When the lethal Boxing Day Tsunami hit the shores of Sri Lanka in 2004, many animals had already fled the scene. It's thought they may have heard the low, infrasonic frequencies of the quake before the tsunami made landfall and thus made their escape. No one knows for sure, but we do know that many animals can hear and communicate in the infrasonic range.

Sound waves and frequency

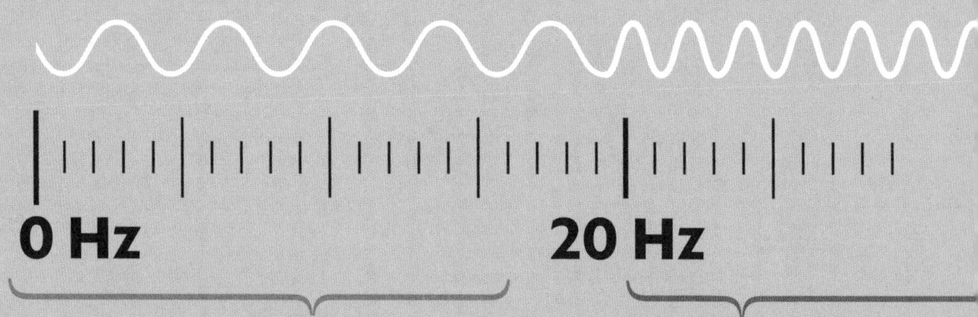

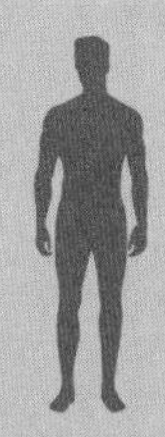

Whales, elephants, hippos, giraffes, and alligators all communicate via infrasound. To attract mates, tigers use low, booming infrasounds and drive away rivals. Elephant calls can travel for several miles, while their foot stomps, which produce infrasonic seismic waves that rumble through the ground, can be sensed through the feet of herds more than 6 miles (10 km) away. It helps members of an otherwise dispersed herd to stay in touch. In water, infrasound travels even farther. A humpback whale singing in the Caribbean can be heard by a fellow whale more than 3,700 miles (6,000 km) away off the west coast of Ireland.

ELEPHANTS "HEAR" WITH THEIR FEET

Pressure receptors, clustered toward the front and back of the feet, may help elephants to detect underground seismic waves, such as those produced by other elephants or even earthquakes. Elephants often lean forward on their toes, or back on their heels, in response to a seismic signal, placing the pressure receptors in closer proximity to the ground. A fatty cushion in the heel of the foot is also thought to help with this.

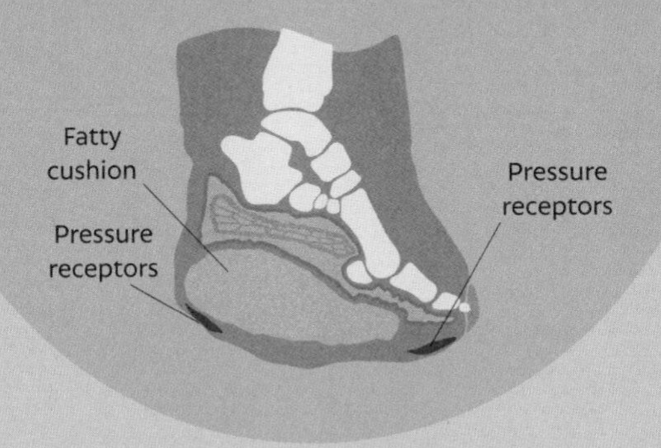

LIFE IN ULTRASOUND

At the other end of the sound spectrum, many animals communicate using high-frequency ultrasound. Bats are perhaps the best-known proponents. They emit pulses of ultrasound that bounce back from objects in the environment. The bats listen for these echoes and use them to figure out the size, shape, and distance of the objects.

Echolocation is used by bats to catch moths on the wing, so moths have evolved some clever counteroffensives. The furry thorax of the Madagascar bullseye moth can absorb up to 85 percent of incoming ultrasound, helping to camouflage the insect from its predators. Some moths simply drop out of the sky when they hear a bat's sonar, while others, like the tiger moth, emit a retaliatory click that jams the bat's sonar, effectively rendering the moth invisible. Dolphins, toothed whales, swiftlets, and the lesser Madagascar hedgehog tenrec also use echolocation.

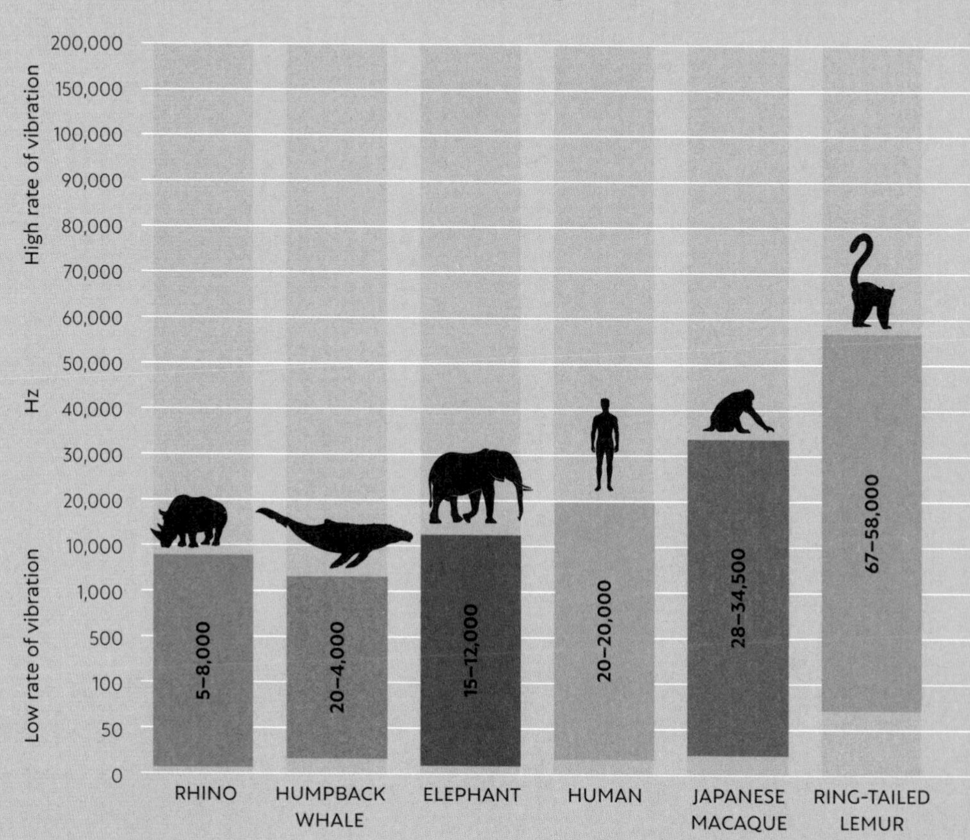

The hearing range of dogs and cats extends into the ultrasound. Dogs can hear sounds of up to 45 kHz (kilohertz), while cats can hear sounds of up to 79 kHz. Dog whistles, used for training, operate at ultrasonic frequencies of 23–54 kHz, which are above the range of human hearing. This explains why dogs can hear them, but we cannot.

The concave-eared torrent frog uses ultrasound to help it find a mate. Most frogs have thick eardrums close to the surface of the skin and can only hear frequencies below 12 kHz. The concave-eared torrent frog, however, has thin, recessed eardrums that enable them to detect higher-pitched frequencies. Females have been recorded chirping at a frequency of around 128 kHz, which helps them to be heard over the noise of the fast-flowing streams where they live in China's Huangshan Hot Springs.

Some frogs have specialized eardrums that help them hear high frequencies.

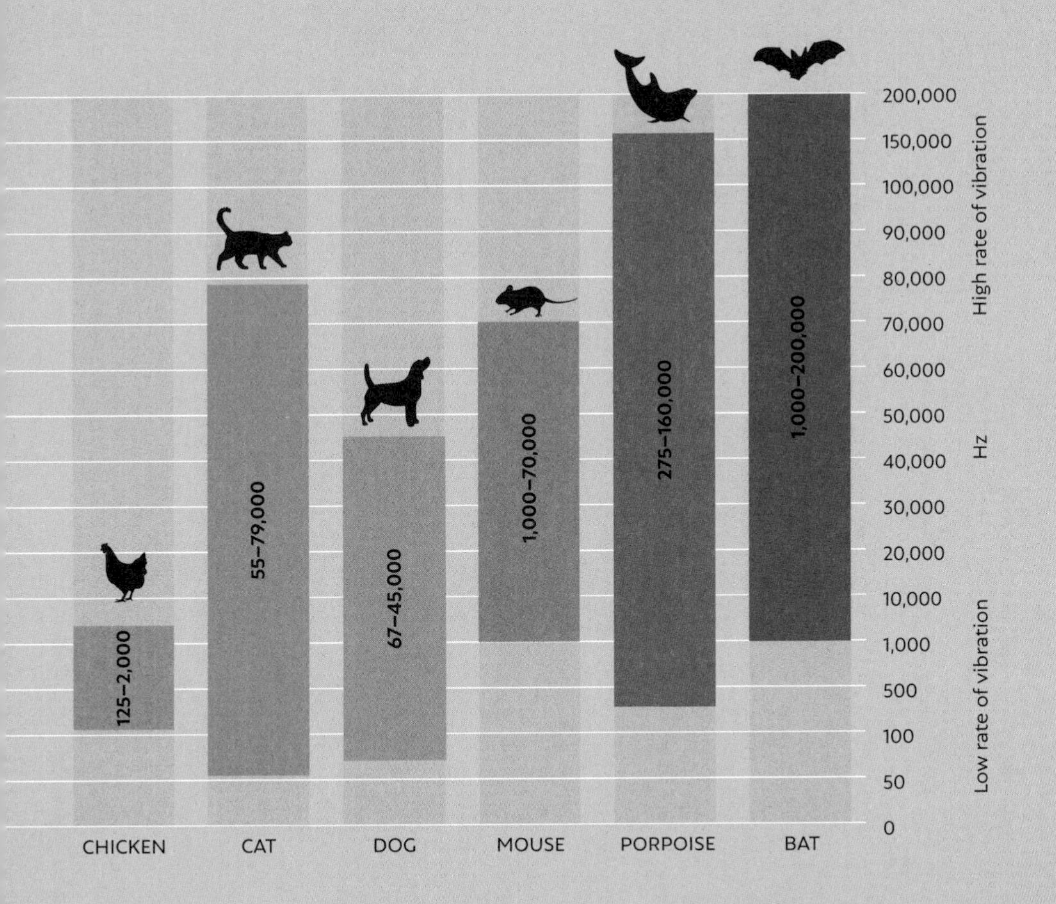

FURTHER READING

BOOKS

Brusatte, Steve. *The Rise and Fall of the Dinosaurs: The Untold Story of a Lost World.* (Picador, London, 2019)

Gow, Derek. *Bringing Back the Beaver: The Story of One Man's Quest to Rewild Britain's Waterways.* (Chelsea Green Publishing, Danvers, 2020)

Kolbert, Elizabeth. *The Sixth Extinction: An Unnatural History.* (Bloomsbury Publishing, London, 2014)

Lloyd, Christopher & Forshaw, Andy. *The Nature Timeline Wallbook (What on Earth Wallbook).* (What on Earth Publishing Ltd, Maidstone, 2017)

Macdonald, Benedict. *Cornerstones: Wild Forces That Can Change Our World.* (Bloomsbury Wildlife, London, 2022)

Monbiot, George. *Feral: Searching for Enchantment on the Frontiers of Rewilding.* (Penguin, London, 2013)

Pilcher, Helen. *Life Changing: How Humans Are Altering Life on Earth.* (Bloomsbury Publishing, London, 2020)

Roberts, Alice. *Tamed: Ten Species That Changed Our World.* (Hutchinson, London, 2017)

JOURNALS

Blanchard, T. S. *et al.* 2019. Phenotypic flexibility in respiratory traits is associated with improved aerial respiration in an amphibious fish out of water. *Journal of Experimental Biology* 222, jeb186486.

Brown, R. B. & Brown, M. B. 2013. Where has all the road kill gone? *Current Biology* 23(6): R233–R234.

Ellis, S. *et al.* 2018. Analyses of ovarian activity reveal repeated evolution of post-reproductive lifespans in toothed whales. *Scientific Reports* 8: 12833.

Gleckler, P. J. *et al.* Industrial-era global ocean heat uptake doubles in recent decades. *Nature Climate Change* 6: 394–398.

Healy, K. *et al.* 2013. Metabolic rate and body size are linked with perception of temporal information. *Animal Behaviour* 86: 4 685–696.

Law, A. *et al.* 2017. Using ecosystem engineers as tools in habitat restoration and rewilding: beaver and wetlands. *Science of the Total Environment* 605–606, 1021–1030.

More, H. L. & Donelan, J. M. 2018. Scaling of sensorimotor delays in terrestrial mammals. *Proceedings of the Royal Society B* 285(1885): 20180613.

O'Connell-Rodwell, C. E. 2007. Keeping an 'ear' to the ground: Seismic communication in elephants. *Physiology* 22(4): 287–294.

Poppinga, S. *et al.* 2012. Catapulting tentacles in a sticky carnivorous plant. *PLoS ONE* 7(9): e45735.

Reppert, S. M. 2007. The ancestral circadian clock of monarch butterflies: Role in time-compensated sun compass orientation. *Cold Spring Harbor Symposia on Quantitative Biology* LXXII: 113–118.

Santini, B. A. & Martorell, C. 2013. Does retained-seed priming drive the evolution of serotiny in drylands? An assessment using the cactus *Mammillaria hernandezii*. *American Journal of Botany* 100(2): 365–373.

Schorr, G. S. *et al.* 2014. First long-term behavioural records from Cuvier's beaked whales (*Ziphius cavirostris*) reveal record-breaking dives. *PLoS ONE* 9(3): e92633.

Stallmann, R. & Harcourt, A. H. 2006. Size matters: the (negative) allometry of copulatory duration in mammals. *Biological Journal of the Linnean Society* 87(2): 185–193.

Stevens, C. E. & Hume, E. D. 1998. Contributions of microbes in vertebrate gastrointestinal tract to production and conservation of nutrients. *Physiological Reviews* 78(2): 393–427.

Talavera, G. *et al.* 2017. Discovery of mass migration and breeding of the painted lady butterfly *Vanessa cardui* in the Sub-Sahara: the Europe–Africa migration revisited. *Biological Journal of the Linnean Society* 120(2): 274–285.

Therrien, F. *et al.* 2021. Mandibular force profiles and tooth morphology in growth series of *Albertosaurus sarcophagus* and *Gorgosaurus libratus* (Tyrannosauridae: Albertosaurinae) provide evidence for an ontogenic dietary shift in tyrannosaurids. *Canadian Journal of Earth Sciences* 58(9): 812–818.

Trumble, S. J. *et al.* 2018. Baleen whale cortisol levels reveal a physiological response to 20th century whaling. *Nature* 9: 4587.

Viera, W. A. *et al.* 2019. Advancements to the axolotl model for regeneration and aging. *Gerontology* 66(3) 1–11.

White, P. W. *et al.* 2019. Spending at least 120 minutes a week in nature is associated with good health and wellbeing. *Scientific Reports* 13: 9(1) 7730.

Yang, P. J. *et al.* 2017. Hydrodynamics of defecation. *Soft Matter* 13: 4960–4970.

Yang, P. J. *et al.* 2014. Duration of urination does not change with body size. *PNAS* 111(33) 11932–11937.

Zheng, J. *et al.* 2018. Breeding biology and parental care strategy of the little-known Chinese Penduline Tit (*Remiz consorbrinus*). *Journal of Ornithology* 159: 657–666.

KEY REPORTS

Climate Change and Marine Conservation: Supporting management
in a changing environment mccip.org.uk/sites/default/files/2021-07/mccip-sandeels-and-their-availability-as-prey.pdf

Feeling the Heat: The fate of nature beyond 1.5°C of global warming wwf.org.uk/sites/default/files/2021-06/FEELING_THE_HEAT_REPORT.pdf

ORGANIZATIONS

World Wildlife Fund
worldwildlife.org

Butterfly Conservation
butterfly-conservation.org

Svalbard Global Seed Vault
seedvault.no

INDEX

IMAGE CREDITS

The publisher would like to thank the following for permission to reproduce copyrighted material:

Alamy: Marli Wakeling 95B; ImageBroker 124T, 150B; Image Source 101; Minden Pictures 71C, 172T; Nature Picture Library 86B, 88T; WorldFoto 145

Gerald R. Allen/Western Australia Museum 106C

Julia Bartoli/Chantal Abergel/IGS/CNRS/AMU 108C

Doklady Biological Sciences 108L

Science Photo Library: Gerry Bishop 111TR; Expedition to the Deep Slope 2007, NOAA-OE 87B; Thomas Marent/Look at Science 90; Reuters 111BR; Sarit Richerson 111BL; John Serrao 140

Shutterstock: Aldona Griskeviciene 125BL; Angelo Giampiccolo 131; Antonio Galvez Lopez 60–1; ay Ondreicka 104L; AyhanTuranMenekay 137C; BelezaPoy 133; Bene_A 89BC; Binturong-tonoscarpe 105B; Butterfly Hunter 184–5; byvalet 169R; Cavan-Images 153C; Chase Dekker 55; COZ 169L; Creeping Things 70TR; Dan Olsen 125BR; Danita Delimont 163R; David Havel 9; Dennis Jacobsen 150C; diegooscar01 107L; Dotted Yeti 124C; Double Brow Imagery 151B; Ecopix 173T; Edwin Godinho 100; Ekaterina Gerasimchuk 124B; Eric Isselee 106L, 153B; Erik Mandre 104R; Etienne Outram 168R; Evelyn D. Harrison 123C; Facanv 193; Grzegorz Dlugosz 164; Hugh Lansdown 71TL; Igor Kruglikov 151T; irinaroma 95T; Israel Moran 115; Jiri Balek 163L; Joshua Davenport 181; Joule Sorubou 163C; Ken Griffiths 71TR; Kerry Hargrove 78T; Kevin Wells Photography 123B; Kirk Wester 151C; komkrit Preechachanwate 127B; Kuttelvaserova Stuchelova 153TC; kwhw 125CL; LeoDeKol 111TL; lucacavallari 89L; Marcin Kadziolka 110; margo_black 179; Mark Time Author 186; Maryna Pleshkun 107B; Matis75 35; MLArduengo 71 BR; mnoor 162C;MZPHOTO.CZ 104C; Peterson 89TR; Phuong-Thao 111CR; Primi2 150T; Ricardo.Flores 125TR; Rob Jansen 67; Ronald Shimek 130; Sallye 174; Salmeroncasanova 91L; shymar2 89R; slowmotiongli 162R; Sue Leonard Photography 78B; Super Prin 149; teekayu 125T; Tom Reichner 70BR; tristan tan 153T; Ua_Biologist 109C; Vadim Petrakov 21; vagabond54 105C; Vagabondering Andy 162L; Vera Larina 153BC; Victoria Tucholka 107T; Vince Adam 201; Vladimir Wrangel 88T; wisawa222 119; wonderisland 89TL; Wonderly Imaging 123T; Zbigniew Guzowski 137T

Wikimedia Commons: 88 BR; Aaron Lucas 105C; Dr Alex Hyatt, CSIRO 70TL; CactiLegacy 161; David V. Raju 95B; Dcrjsr 85; Diego Fontaneto 108L; Guillaume Dargaud 87T; Jiang Chunsheng/ doi.org/10.1038/s41598-021-92372-z 66; © 2011 Jee & Rani Nature Photography 173B; Jim Rorabaugh/ USFWS 71BL; Karl Brodowsky 91R; Kembangraps 168L; Leyo 86T; Nico Michiels 172B; NOAA Okeanos Explorer Program 87C; Oskar Liset Pryds Hansen 109L; Petr Hamerník 70 BL; S. Rae 86C; Steve Trewick 109R; Taollan82, Kirt L. Onthank 88BC; USDA Forest Service 79; writings.stephenwolfram.com/2018/01/showing-off-to-the-universe-beacons-for-the-afterlife-of-our-civilization 169L

The publishers would like to acknowledge the following sources:

p. 28 en.wikipedia.org/wiki/Evolution_of_the_horse#/media/File:Horseevolution
p. 56 marinesanctuary.org/wp-content/uploads/2021/01/whale-fall-poster-noaa-onms
p. 74 news.mongabay.com/2015/06/study-confirms-what-scientists-have-been-saying-for-decades-the-sixth-mass-extinction-is-real-and-caused-by-us
p. 80 oregonzoo.org/sites/default/files/gallery/images/Condor_graphic_web_H
p. 138 sitn.hms.harvard.edu/flash/2018/regeneration-axolotl-can-teach-us-regrowing-human-limbs
p. 140 rzuser.uni-heidelberg.de/~bu6/Introduction03.html
p. 180 learning.rzss.org.uk/mod/book/view.php?id=1277&chapterid=668

Every effort has been made to trace copyright holders and to obtain their permission for the use of copyright material. The publisher apologizes for any errors or omissions in the lists above and will gratefully incorporate any corrections in future reprints if notified.